The Bioarchaeology of Children

This book is the first to be devoted entirely to the study of children's skeletons from archaeological and forensic contexts. It provides an extensive review of the osteological methods and theoretical concepts of their analysis. Non-adult skeletons provide a wealth of information on the physical and social life of the child from their growth, diet and age at death, to factors that expose them to trauma and disease at different stages of their lives. This book covers non-adult skeletal preservation; the assessment of age, sex and ancestry; growth and development; infant and child mortality including infanticide; weaning ages and diseases of dietary deficiency; skeletal pathology; personal identification; and exposure to trauma from birth injuries, accidents and child abuse, providing new insights for undergraduates and postgraduates in osteology, palaeopathology and forensic anthropology.

MARY E. LEWIS is a lecturer at the University of Reading and has taught palaeopathology and forensic anthropology to undergraduate and postgraduate students for over 10 years. Mary is also an advisor to the police and has served as a registered forensic anthropologist for the Ministry of Defence.

Cambridge Studies in Biological and Evolutionary Anthropology

Series editors

HUMAN ECOLOGY
C. G. Nicholas Mascie-Taylor, University of Cambridge
Michael A. Little, State University of New York, Binghamton
GENETICS
Kenneth M. Weiss, Pennsylvania State University
HUMAN EVOLUTION
Robert A. Foley, University of Cambridge
Nina G. Jablonski, California Academy of Science
PRIMATOLOGY
Karen B. Strier, University of Wisconsin, Madison

Also available in the series

The Bioarchaeology of Children

Perspectives from Biological and Forensic Anthropology

Mary E. Lewis
University of Reading

CAMBRIDGE
UNIVERSITY PRESS

CAMBRIDGE UNIVERSITY PRESS
Cambridge, New York, Melbourne, Madrid, Cape Town, Singapore,
São Paulo, Delhi, Dubai, Tokyo

Cambridge University Press
The Edinburgh Building, Cambridge CB2 8RU, UK

Published in the United States of America by Cambridge University Press, New York

www.cambridge.org
Information on this title: www.cambridge.org/9780521121873

First published 2007
This digitally printed version 2009

A catalogue record for this publication is available from the British Library

ISBN 978-0-521-83602-9 Hardback
ISBN 978-0-521-12187-3 Paperback

Contents

Acknowledgements

Many friends and colleagues have provided advice and encouragement in the preparation and writing of this book. In particular, Charlotte Roberts, Keith Manchester and Jenny Wakely were instrumental in first introducing me to human remains, and Charlotte Roberts has continued to provide invaluable advice and support throughout my career. I am indebted to many who provided unpublished data, assistance, access to images and comments on drafts of the text: Kristine Watts, Anthea Boylson, Louise Loe, Hella Eckardt, Louise Humphrey, Roberta Gilchrist, Margaret Cox, Rebecca Gowland, Rebecca Redfern, Gundula Müldner, Luis Rios, Bill White, Donald Ortner and Richard Steckle. I also thank the team at Cambridge University Press, Tracey Sanderson, Dominic Lewis, Emma Pearce and Anna Hodson.

I am thankful to all my friends and colleagues at the Department of Archaeology in Reading for making it such a stimulating place to work and who have allowed me the time and funds needed to finish this book. Finally, I am grateful to my family for their continued support.

Figures 6.4, 8.4 and 8.5 are from the research slide collection of D. J. Ortner, Department of Anthropology, Smithsonian Institution, Washington, DC, digitised and made available through funds supporting National Science Foundation grant SES-0138129 by R. H. Steckel, C. S. Larsen, P. W. Sculli and P. L. Walker (2002), *A History of Health in Europe from the Late Paleolithic Era to the Present* (Mimeo, Columbus, OH). These images are reprinted with their kind permission. Figures 6.9–6.11, 7.1, 7.2 and 8.6 are of specimens from the Human Remains Collection, Biological Anthropology Research Centre, Department of Archaeological Sciences, University of Bradford. Many of these photographs were taken by Jean Brown, to whom I am very grateful. Every effort has been made to acknowledge copyright holders, but in a few cases this has not been possible. Any omissions brought to my attention will be remedied in future editions.

1   *The bioarchaeology of children*

1.1   Children in archaeology

This book reviews the current status of children's skeletal remains in biological and forensic anthropology. Child skeletons provide a wealth of information on their physical and social life from their growth and development, diet and age at death, to the social and economic factors that expose them to trauma and disease at different stages of their brief lives. Cultural attitudes dictate where and how infants and children are buried, when they assume their gender identity, whether they are exposed to physical abuse, and at what age they are considered adults. Similarly, children may enter the forensic record as the result of warfare, neglect, abuse, murder, accident or suicide and the presence of young children within a mass grave has powerful legal connotations. The death of a child under suspicious circumstances is highly emotive and often creates intense media coverage and public concern, making the recovery and identification of their remains more pressing. In forensic anthropology, techniques used to provide a biological and personal identification as well as the cause and manner of death provide particular challenges.

The study of children and childhood in social archaeology emerged out of gender theory in the 1990s, and has gradually increased in its sophistication, moving children out of the realm of women's work, to participating and active agents in the past, with their own social identity, material culture and influence on the physical environment around them. Children who were once invisible in the archaeological record are slowly coming into view. The primary data for the archaeology of childhood are the children themselves, and in order to progress this new discipline, it is important to examine how bioarchaeologists derive the data from which social interpretations are made, and the limitations that are inherent in the methods and nature of immature skeletal material, including the impact of the burial environment on their recovery.

Comparative studies of children from archaeological contexts have been complicated by the eclectic use of terminology that both describes the skeleton as a child and prescribes an age for the individual. For example, the use of the term 'infant' properly assigned to those under 1 year of age, has been used to describe children aged up to 5 years, whereas 'juvenile' can be divided into 'juvenile I'

1

Table 1.1 *Age terminology used in this volume*

Term	Period
Embryo	First 8 weeks of intra-uterine life
Fetus	From 8 weeks of intra-uterine life to birth
Stillbirth	Infant born dead after 28 weeks gestation
Perinatal, perinate	Around birth, from 24 weeks gestation to 7 postnatal days
Neonatal, neonate	Birth to 27 postnatal days
Post-neonatal	28–346 postnatal days (1 year)
Infant	Birth to 1 year
Non-adult	≤17 years
Child	1–14.6 years
Adolescent	14.6–17.0 years
Adult	>17 years

or 'juvenile II' with a variety of ages assigned. One of the most popular terms used by osteologists to describe children is 'sub-adult'. This term is problematic as it has been used to define a specific age category within the childhood period. More fundamentally, sub-adult implies that the study of these remains is somehow less important than that of the adults (i.e. sub = below). Throughout this book children are described as 'non-adults' encompassing all children recovered from the archaeological record up to the age of 17 years. Additional terms divide this overarching category into critical physiological periods of the child's life (Table 1.1). These terms are used for ease of reference and provide a biological basis for discussion; they are not intended to describe the complex social experience of the youngest members of every society, past or present.

This book is divided into nine chapters, covering the development of childhood archaeology and the osteological study of non-adult remains; factors affecting preservation; assessment of their age, sex and ancestry; growth and development; infant and child mortality including infanticide; weaning ages and diseases of dietary deficiency; skeletal pathology; and exposure to trauma from birth injuries, accidents and child abuse. The final chapter considers some future directions for the study of children in bioarchaeology. The following sections explore the gradual development of childhood theory in archaeology and the rise of research into non-adult skeletal remains in both biological and forensic anthropology.

1.2   A history of childhood

Studies of the history of childhood began in 1960 when Philip Ariès published *Centuries of Childhood: A Social History of Family Life*. Ariès argued that the 'childhood' we know today, which may perhaps be described as a period of 'cosseted dependency' (Derevenski, 2000:4), did not exist until the early modern period. Prior to this, parents were unsympathetic and detached from their children, dressing them and expecting them to behave as miniature adults. Such indifference was considered a coping mechanism to the constant threat of infant mortality (Ariès, 1960). In the past, we were led to believe, a child's upbringing was a combination of neglect and cruelty. Further debates in the 1970s developed the theme (De Mause, 1974; Shorter, 1976; Stone, 1977), while later discourses began to challenge this traditional view (Attreed, 1983; Hanawalt, 1986, 1993; Swanson, 1990; Shahar, 1992). Historians and social archaeologists have now updated and revised our impressions of childhood. In past societies, stages of life that correspond to childhood were recognised and marked by social events or burial practices. Many parents loved their children, sometimes to distraction. For example, Finucane (1997) concentrated on the 'miracle' texts of the medieval period which contained numerous tales of family and village reactions to a child's death or illness, with parents crippled by grief or friends and relatives praying by a riverbank for the recovery of a drowned child. Although important, these studies focussed on the attitude of adults towards children, rather than viewing the past through a child's eyes.

The study of children and childhood in archaeology emerged out of gender theory in the 1990s (Derevenski, 1994, 1997; Moore and Scott, 1997). Previously, children had been considered 'invisible' in the archaeological record, but a feminist reassessment of the past placed specific emphasis on gender and age and with this, on the nature of childhood. Lillehammer (1989) was one of the first to address the role of children in archaeology. She suggested that through the use of burial, artefacts, ethnography and osteology we could gain insight into the relationship the child had both with its physical environment and the adult world. This was followed by an examination of documentary and archaeological evidence for the child in the Anglo-Saxon and medieval periods (Coulon, 1994; Crawford, 1999; Orme, 2001), with Scott (1999) providing a multicultural view on aspects of infancy and infanticide. Crawford (1991) studied the Anglo-Saxon literature for clues as to when children were subject to adult laws. Beausang (2000) expanded this theory of childhood to incorporate the concepts and practice of childbirth in the past, with the recognition of birthing artefacts in the archaeological record. Although a promising start, these studies have been criticised for maintaining the idea that children were passive recipients in their communities, invariably linked to the activities of women (Wilkie, 2000). Furthermore, the category of 'child' is often used in order to investigate the construction of 'adult' (Derevenski, 2000). Neither approach allows us to explore the role of the child as an independent agent in the past. Wilkie (2000) went some way to redress this balance when she used evidence of the toy industry in the eighteenth and nineteenth centuries to illustrate how, through their own material culture, children displayed their sense of identity and defined their own distinctive social networks and liaisons.

1.2.1   Defining childhood

> BOREDOM !!! SHOOTING!!! SHELLING!!! PEOPLE BEING
> KILLED!!! DESPAIR!!! HUNGER!!! MISERY!!! FEAR!!! That's my
> life! The life of an innocent eleven-year-old schoolgirl!! ... A child without
> games, without friends, without sun, without birds, without nature, without
> fruit, without chocolate or sweets ... In short, a child without a childhood.
>> Extract from the diary of a child in the Sarajevo conflict, 1992; from Cunningham
>> (1995:1)

As this entry from the diary of a child in war-torn Sarajevo testifies, children have an expectation of what childhood should be. No matter what period we are examining, childhood is more than a biological age, but a series of social and cultural events and experiences that make up a child's life. Childhood can be defined as a period of socialising and education, where children learn about their society, gender roles and labour through play. The initial dependence on their parents for nourishment and protection slowly diminishes as the child ages and becomes an independent member of society. The time at which these transitions take place varies from one culture to another, and has a bearing on the level of interaction children have with their environment, their exposure to disease and trauma, and their contribution to the economic status of their family and society. The Western view of childhood, where children do not commit violence and are asexual, has been challenged by studies of children that show them learning to use weapons or being depicted in sexual poses (Derevenski, 2000; Meskell, 2000). What is clear is that we cannot simply transpose our view of childhood directly onto the past.

Bogin (1997, 1998) takes an evolutionary approach to childhood theory. Childhood is a period in the human life cycle not found in any other mammal, and for Bogin this is defined as a period of time between the ages of 3 and 7 years, when 'the youngster is weaned from nursing but still depends on older people for feeding and protection' (Bogin, 1997:64). The child is constrained by its immature dentition, small digestive system and calorie-demanding brain, which influence the type and amounts of food it can consume. 'Juvenility' occurs with the eruption of the permanent dentition, and when children are able to procure and consume their own foods, as the brain and body growth diminish to less than 50% of total energy needs, and they undergo a cognitive shift. This period begins at the age of 7 and ends with the onset of puberty (*c.*10 years in girls, *c.*12 years in boys). Bogin (1998) asserts that in humans, childhood performs several functions: an extended period for brain growth, time to acquire technical skills, time for socialisation and an aid to adult reproduction. That is, that the childhood period allows the mother to wean the child and produce other offspring, by passing the energy expenditure of feeding and caring for the child onto siblings and post-reproductive members of society, such as grandparents (Key, 2000). This urge to care for the child is manipulated through the child's retention of its infantile appearance (large cranium, small face and body); that is to say, children are 'cute'. As the body and brain slow in their growth during this period, they require less energy expenditure to feed but are protected during times of hardship (Bogin, 1998). Many would object to this purely biological view of childhood, as it ignores social theories of when a child becomes an 'adult' and a fully fledged member of a society, something that is culturally defined. Hanawalt (2002) argues that in order for a child to survive, it must not only be nursed, fed and kept warm (biological survival), but also be played with and talked to (cultural survival).

1.2.2 *Defining the child: biological versus cultural age*

One of the resounding issues with the definition of a 'child' in archaeological contexts is the use of physiological age to determine a social category (Gowland, 2001; Baxter, 2005). Physiological age is a biological reality, whereas 'child' is a culturally loaded term. The age at which an individual leaves the world of dependency, learning and play, and takes on roles of work and social responsibility is neither distinct nor universal. That there are three types of age category, 'biological', 'chronological' and 'social', is not denied, but in order to examine the past life-course we need to have consistency in the raw data (the skeletal remains), and use accurate osteological assessments of age and physiological development as a marker from which to base our interpretations of the social understanding of age in the past. Biological age is not irrelevant in the way in which society treats a child. It affects children's connection to their physical and social environment, from total dependency during infancy, to when they begin to crawl, walk, talk and communicate with the adults and children around them (Table 1.2). These abilities are physiologically determined and they dictate how the child interacts. In particular, the misuse of the term 'infant' to refer to children between the ages of 1 and 3 years or 1 and 5 years in studies that use skeletal evidence as their data misses this point. As an infant (under 1 year), the child is particularly vulnerable to disease and death, and its chances of survival significantly increase after the first year. Children who die at around 2 years of age may be reflecting inadequate weaning methods or

Table 1.2 *Child development milestones from birth to 5 years*

Birth to 8 months	8 months to 1.5 years	1.5 to 3 years	5 years
Lifts and holds up head	Begins to crawl and may stand aided by furniture	Stands on one foot or on tiptoe	
Turns over unaided (7 months)	Can throw without losing balance	Can run, skip, climb and has a developed sense of balance	Dresses and undresses
Reaches towards objects	Handles finger-foods Uses spoons and cups	Imitates others	
	Becomes anxious when separated from loved ones	Understands people and objects still exist when they cannot be seen	
Smiling and gazing	Shows affection by kissing and hugging	Expresses pride, pleasure, embarrassment and shame	
	Responds to name	Listens to stories	Tells stories
	Explores environment	Understands the future and the past	
	Interacts with other children		Social interaction and role-playing
Gurgles and babbles to communicate	Forms simple sentences	Uses sentences to communicate feelings and needs	Asks questions about the meaning of words
	Has no understanding of 'male' and 'female'	Understands 'male' and 'female' through dress and over time, but not changing situations	Understands 'male' and 'female' through time and situations: 'gender consistency'

Source: Collated from Berhrman *et al.* (1996) and Kohlberg (1966).

unsanitary conditions, and those that make it to 3 years are talking, playing and actively mobile. By 5 years they are capable of contributing to the household with minor chores. To categorise this most vital developmental period into one age category, 'infant', will mask important physiological and, hence, social advancements.

   Derevenski (1997) refers to Kohlberg's (1966) work on a child's understanding of gender roles. Before the age of 2, a child has no concept of male or female but after 2 years of age, they begin to recognise males and females by visual prompts such as clothing. Between the ages of 3 and 4 years, a child's concept of gender becomes stable, and is understood through time. Hence, if you are male when you are young, the child understands that you will be male as an adult, but if a male begins to perform what the child perceives as female roles, the male would become female. A stage of 'gender consistency' through time and situation is not reached until the child is 5 (Table 1.2). Wiley and Pike (1998) suggested the use of developmental stages rather than chronological age to devise child mortality rates to take into account the activity of the child (crawling, weaning, walking), which is often related to their cause of death through exposure to disease and accidental injury. Although they propose this method for use in modern communities where calendar age is rarely recorded, the application of such developmental age categories into archaeological studies has the advantage of placing the child at the centre of the study by examining the environment from their vantage point.

Although biological age categories provide data from which interpretations are made, adult perceptions of the ability, maturity and responsibilities of children at each age are culturally determined, and must be considered when trying to ask questions about past child activity and health. In the later medieval period, the ages of 8–12 years represented a time when children would begin their apprenticeships (Cunningham, 1995), and children as young as 12 and 14 years could be married in ancient Egypt and Rome respectively, leaving the realm of child for that of wife and mother. Childbirth is not a common interpretation for the cause of death for older children within the burial record. Today in the UK, children reach adulthood by degrees. At 16 they can legally have sex, at 17 they can learn to drive, at 18 they can drink, get married and vote, reflecting their status as full members of society. Crawford (1991) rightly criticises archaeologists for their inconsistency in choosing the cut-off point for children in archaeological reports, which vary from 15 years to 25 years in some cases. These inconsistencies have a great impact on the way in which a cemetery is interpreted. Moving an individual from one age category to another can fundamentally change the profile of a cemetery when attempting to evaluate the pattern of adult and non-adult burials, and to understand the significance of their grave inclusions.

Attempts to define periods of transition in childhood have been carried out by examining the burial of children and the engendered nature of their gravegoods at certain ages. Gowland (2001, 2002) noted that at Romano-British Lankhills in Hampshire, children were buried with gravegoods from the age of 4 and the quantity of artefacts peaked between 8 and 12 years. Gowland (2001) suggests that in these communities at least, age thresholds appear at infancy (where perinates are interred outside of the cemetery area), at 4 years and between 8 and 12 years where the quantity and wealth of gravegoods increases. Stoodley (2000) examined the presence of certain gravegoods within burials from a large number of Anglo-Saxon graves in England. He noted that 'masculine' spears began to appear more frequently in male graves after the age of 10–14 years, whereas 'female' beads and dress adornments appeared in 'girl's' graves at between 10 and 12 years. This study suffers from a common circular argument which stems from our inability to provide a biological sex for non-adults, and a Westernised view of what is 'masculine' and what is 'feminine'. This circle was partially broken in Rega's (1997) study of burials from Bronze Age Mokrin in Yugoslavia, where children were sexed using canine tooth-crown dimensions. Using these data, Rega revealed that all children were provided with the same feminine engendered artefacts found in adult female graves until around 17 years of age, when individuals sexed as male began to be buried with artefacts associated with the male adult graves. Stoodley's (2000) age bracket in the Anglo-Saxon childhood life-course is supported by Crawford's (1991) analysis of contemporary records revealing that children as young as 10 years could inherit property and be prosecuted under adult laws. Kamp (2001) provides an excellent review of the development of childhood studies and argues that the age categories employed by osteologists are often selected and compared without reference to the society in which the children lived. Biological or physical development and social markers of childhood are not always related. This was demonstrated in Van Gennep's (1960) *The Rites of Passage* in which physical puberty did not always coincide with the rites of passage that marked the adolescents' entry into the adult world. Archaeological evidence from the Anglo-Saxon period also attests to this, with male adult-type gravegoods only appearing once an individual has reached 20–25 years (Stoodley, 2000), some 6 years after they would have reached puberty. While the study of childhood has come some way in elucidating a particular section of the human life-course, Gilchrist (2004) calls attention to the fact that other age categories are still neglected, among them, what it was to be an adolescent in the past.

1.2.3   Children in the archaeological record

Some artefacts have provided tangible links to children in the past. Footprints (Roveland, 2000), death masks (Coulon, 1994), fingerprints on pots (Baart, 1990) and tooth marks in resin (Aveling, 1997) all prove that a child was there. Wilkie's (2000) discussion of toys that were designed, manufactured and sold with children in mind forced historical archaeologists to acknowledge them as actors in past society, but this concept has been slow to catch on in time periods where the material evidence is not so rich. It may be that our association with children and toys is based on Western ideals of what childhood should be, and

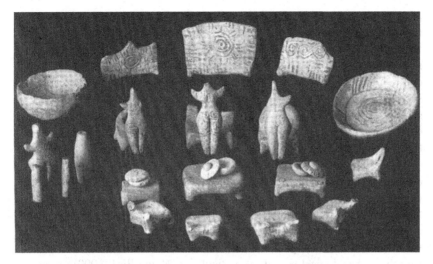

Figure 1.1   Possible toys from the Ovcarovo 'cult scene'. From Whittle (1996:94), reproduced with kind permission from Cambridge University Press.

this has led some scholars to avoid toys as a route to the activities of children (Derevenski, 1994). Nevertheless, humans learn through play, trial and error and it is conceivable that small items or badly drawn or sculpted figures in the archaeological record were used and created by children. Just as female engendered space is now recognised in the past, it is time to start considering the potential of identifying childhood spaces, where 'women and children' are no longer seen as one entity and children are viewed as independent agents within their own social space (Wilkie, 2000). Children have the imagination to make toys out of sticks, stones and everyday household objects that will be invisible in the archaeological record. In this way, children may influence the formation processes of a site, perhaps by the movement of artefacts from their original site of deposition (e.g. a midden), and the physical alteration of household objects. A small pile of stones or an unusual collection of post-holes may indicate a child was at play, and this possibility should be taken into account when interpreting a site. Until recently, child activity in the archaeological record has been seen as detracting from the real issues of adult behaviour (Bonnichsen, 1973; Hammond and Hammond, 1981), rather than being viewed as informative of the child's interaction with its physical environment.

Possible toys have been recovered from various sites throughout Europe. Of particular note are the small decorated clay figures, miniature furniture and tiny bowls found at Ovčarovo, Bulgaria (Fig. 1.1), and the clay house and figurines located in a house at Platia Magoula Zarkou, northern Greece, both finds dating to the Neolithic (Whittle, 1996). Rossi (1993) identified two ivory dolls in the grave of a Roman child from Yverdon-les-Bains, Switzerland. Such items were traditionally interpreted as 'cult' objects or foundation offerings, rather than as a child's playthings. On the other hand, the idea that all miniaturised items represent toys is overly simplistic. Sillar (1994) noted that in the Andes, while children will play with miniature pots, mimicking adult household practices such as cooking and trade, such pots were also used by adults as donations at shrines. In lithics analysis, small cores have been interpreted as being made by children mimicking the adult knappers. Finlay (1997) suggests that inconsistently made lithic artefacts may be the work of young apprentices, learning the trade and that, as producers, children would make lithics in keeping with the adult norms, rather than on a miniature scale. Bird and Bird (2000:462) argue that differences between adult and child foraging patterns are not always about the learning process, and that 'children are not always practicing to be good adults . . . but are predictably behaving in ways that efficiently solve immediate fitness trade-offs'. If this pattern is predictable then we should be able to identify it in the archaeological record. In particular, Bird and Bird (2000) examined the different adult and child patterns of shellfishing in the Eastern Torres Strait on the Great Barrier Reef. Due to their inexperience, children tended to collect a wider variety of less valuable shellfish, which they proceeded to eat, leaving them in small middens outside the settlement. Adults were able to exclusively collect the most profitable and difficult-to-gather shellfish, avoiding the types the children collected. In the archaeological record, two forms of shell midden in different locations should be evident, with the more diverse and marginal middens representing the foraging patterns of the children.

1.3   Children in biological anthropology

The study of children in biological anthropology has earlier beginnings than in social archaeology, but they were no less focussed. Most studies were stimulated by an interest in fertility levels, or the information that child survival could provide on adult adaptation to their changing surroundings. These endeavours were constantly being frustrated by the perceived notion that infant and child remains could not survive the burial environment. It was only in the 1990s that the study of non-adult skeletons began to concentrate on the information that could be provided on the growth and health of the children themselves, providing information on their activities and risk of infection or injury in contrasting environments. Examination of the physical remains of children provides us with the most direct and intimate evidence for them in the past. This section outlines the development of the study of child skeletal remains in biological anthropology and palaeopathology up until the present day.

Before the 1980s, studies of non-adult skeletal remains concentrated on devising ageing and sexing methods based on medical studies (e.g. Schour and Massler, 1941; Hunt and Gleiser, 1955). For example, Balthazard and Dervieux (1921), Scammon and Calkins (1923) and later Olivier and Pineau (1960) provided data for fetal ageing using diaphyseal lengths, while Boucher (1955, 1957) assessed the use of the sciatic notch for sexing infant remains. In the 1960s studies on the physical growth of past populations began to emerge, and would dominate research in non-adult remains for the next 40 years. The most prolific researcher in this area at the time was Francis Johnston, who examined the growth of children from Indian Knoll in Kentucky (Johnston and Snow, 1961; Johnston, 1962, 1968). Following Johnston's example, by the late 1970s the majority of studies that included child remains were focussed on diaphyseal length measurements to estimate growth attainment (Armelagos *et al.*, 1972; Y'Edynak, 1976; Merchant and Ubelaker, 1977). With the increasing interest in palaeodemography, researchers began to assess the impact of under-representation of child remains on life tables (Moore *et al.*, 1975), but only a few were interested in what these data could contribute to our understanding of perinatal and child mortality (Brothwell, 1971; Henneberg, 1977; Mulinski, 1976). In palaeopathology, iron-deficiency anaemia as the underlying cause of cranial porous lesions (porotic hyperostosis) was under increasing discussion, with several studies examining its prevalence in non-adult crania (El-Najjar, 1977a; Lallo *et al.*, 1977). However, the association of enamel hypoplasias and Harris lines with childhood stress was indirectly determined using adult skeletal and dental material (McHenry, 1968; McHenry and Schulz, 1976; Rose *et al.*, 1978). In 1978, Mensforth and colleagues heralded a way forward when they examined the prevalence of anaemia and infection (i.e. porotic hyperostosis, periostitis and endocranial lesions) in 452 infants and children from the Late Woodland ossuary sample from the Libben site in Ottawa County, Ohio. For the first time, the health of children in the past was the primary focus of study (Mensforth *et al.*, 1978). This research also demonstrated the importance of healed and active lesions in determining the precise age at which children were most at risk; the kind of detail not available when using adult evidence. In the same year, Fazekas and Kósa (1978) published their detailed study of Hungarian fetal skeletal remains, raising awareness of the number and morphology of these tiny bones.

In 1980, Buikstra and Cook summed up child studies as being hindered by poor preservation, lack of recovery and small sample sizes, despite, they argued, many researchers becoming aware of their importance in determining the overall success of a population (Buikstra and Cook, 1980). Instead, there was a proliferation of papers on the lack of preservation of non-adult remains compared to adults (Gordon and Buikstra, 1981; Von Endt and Ortner, 1984;

Walker *et al.*, 1988), an assumption that still prevails today. The prevalence of stress indicators in children became more popular as researchers began to assess the impact of agriculture, colonisation and urbanisation on child health (Blakey and Armelagos, 1985; Storey, 1986, 1988). Jantz and Owsley (1984; Owsley and Jantz, 1985) demonstrated changes in child health of the Arikara between AD 1600 and AD 1835 as a result of malnutrition and maternal stress brought about by contact with European settlers. Goodman and Armelagos (1989) highlighted the importance of children under 5 as the most sensitive members of society to environmental and cultural insults, whose stress experience would impact on the overall population's ability to rally from disease in adulthood. Schultz (1984, 1989) began to examine the health of non-adults from around the world, employing histological analysis for evidence of scurvy and tuberculosis. In Egypt, Brahin and Fleming (1982) reported on the health of child skeletal and mummified remains, reporting the presence of tuberculosis, spina bifida and osteogenesis imperfecta, while commenting on the lack of evidence for rickets. Our inability to diagnose rickets in skeletal remains was about to come under scrutiny (Stuart-Macadam, 1988).

By the 1990s, non-adults were becoming incorporated into biocultural studies of different populations (e.g. Stuart-Macadam, 1991; Grauer, 1993; Higgins, 1995; Ribot and Roberts, 1996). These studies were encouraged by the increase in the non-adult material available. Children of known age and sex from Christ Church Spitalfields and St Bride's Church in London became accessible (Molleson and Cox, 1993), and data began to be published on non-adults from Wharram Percy (Mays, 1995) and St Thomas' Church in Belleview, Ontario (Saunders *et al.*, 1993a; 1995). These samples encouraged a revival of methods to estimate the sex of non-adults (De Vito and Saunders, 1990; Mittler and Sheridan, 1992; Schutkowski, 1993; Loth and Henneberg, 1996; Molleson *et al.*, 1998). Saunders (1992) carried out a review of non-adult growth studies, outlining their advantages and limitations, particularly the issue of comparing deceased children to living healthy modern populations (Saunders and Hoppa, 1993), while others began to highlight the potential and extent of pathological evidence that could be derived from their study (Anderson and Carter, 1994, 1995; Lewis and Roberts, 1997). Nearly 10 years after Stuart-Macadam (1988) had raised the issue, Ortner and colleagues began to address the diagnosis of rickets and scurvy (Ortner and Ericksen, 1997; Ortner and Mays, 1998; Ortner *et al.*, 1999), while others identified sickle-cell anaemia, juvenile rheumatoid arthritis (Still's disease) and leprosy (Hershkovitz *et al.*, 1997; Rothschild *et al.*, 1997; Lewis, 1998). By the end of the decade, dental microstructure was being used to refine ageing techniques (Huda and Bowman, 1995) and stable isotope analyses to address the age of weaning in contrasting past populations throughout the world were beginning to dominate the literature (Katzenberg and Pfeiffer, 1995; Katzenberg *et al.*, 1996; Schurr, 1997; Wright and Schwartz, 1997; Herring *et al.*, 1998; Wright, 1998).

Today, studies of infant and child skeletal remains are receiving much more attention. The publication of texts on non-adult osteology has increased the number of researchers familiar with their identification and anatomy (Scheuer and Black, 2000, 2004; Baker *et al.*, 2005). Children are routinely included in wide-ranging studies of health in the past (Steckel and Rose, 2002; Cook and Powell, 2005), while the analysis of children themselves from sites all over the world continues (Baker and Wright, 1999; Buckley, 2000; Lewis, 2002a; Bennike *et al.*, 2005; Blom *et al.*, 2005). As we refine our ageing techniques and statistical methods (Gowland and Chamberlain, 2002; Tocheri and Molto, 2002; Fitzgerald and Saunders, 2005) our understanding of the importance of childhood diseases and their diagnosis is becoming more advanced and widely publicised (Glencross and Stuart-Macadam, 2000; Ortner *et al.*, 2001; Santos and Roberts, 2001; Lewis, 2002b, 2004; Piontek and Kozlowski, 2002). New understanding of trauma in the child has meant we can now reassess the evidence for physical abuse and occupational injury, to gain a fuller understanding of the child's life experience in past society.

In biological anthropology, we still wrestle with the issue of children in the archaeological sample representing the 'non-survivors' from any given population. Their pattern of growth or frequency of lesions might not reflect that of the children that went on to survive into adulthood (Wood *et al.*, 1992; Saunders and Hoppa, 1993). The early death of these individuals provides other challenges in the study of non-adult palaeopathology. Chronic diseases need time to develop on the skeleton, but the children that enter the archaeological record have usually died in the acute stages of disease before the skeleton has had time to respond (Lewis, 2000). At the present time, studies that concentrate on non-adult material are hindered by the inability to make reliable sex estimations, due to absence of the secondary sexual characteristics evident on the adult skull and pelvis. Although sexual dimorphism has been identified in utero, there is still a disagreement about the validity of identifying morphological traits indicative of sex in the non-adult skeleton (Saunders, 2000). However, the application of ancient DNA analysis in determining the sex of non-adult skeletal material holds promise for the future.

1.4   Children in forensic anthropology

Forensic anthropology is the application of biological anthropological techniques to the study and identification of skeletal remains recovered from a crime scene. Forensic anthropologists frequently work in conjunction with forensic pathologists and odontologists to suggest the age, sex, ancestry, stature (biological identification) and unique features (personal identification) of the deceased individual from the skeleton. Forensic anthropologists also contribute to the understanding of skeletal trauma to aid in the determination of the cause and manner of death. The data collected from the analysis constitute evidence to be presented in a court of law. Children may enter the forensic record through warfare (e.g. as child soldiers), abuse, murder, accident, suicide or neglect, but the presence of young children within a mass grave has powerful legal connotations and is highly emotive. The death of a child under suspicious circumstances creates intense media coverage and public concern, making the recovery and identification of the remains more pressing and objectivity more difficult to maintain (Lewis and Rutty, 2003).

The biological and personal identification of children's remains in forensic anthropology is hindered by the paucity of techniques usually employed to provide such identification in adults (Kerley, 1976). Features that denote ancestry and sex usually develop after puberty, when hormone levels increase and sexual dimorphism becomes more apparent in the skull and pelvis. For example, racial differences of the mid-facial projection and the appearance of the nasal root develop during puberty, as do brow ridges used in sexing the skull. Estimations of ancestry and sex are crucial to provide an accurate assessment of both age and stature in skeletal remains, as they have an effect on the rate of growth and development (see Chapter 3). An estimation of the minimum number of individuals (MNI) is often easier to obtain in children, as sizes vary with age and between individuals. However, in some cases children of similar age may be recovered and size may not be a useful distinguishing feature, especially where the epiphyses and developing dentition are concerned. Young children seldom visit the dentist, or have major surgery, and their abuse or neglect can hamper the one technique in which anthropologists are most confident when examining child remains: age estimation.

Probably one of the greatest limiting factors in the development of standards for child identification stems from the lack of modern non-adult skeletal collections. Parents rarely choose to donate their children's bodies to medical science, a situation not aided by events in England in 1999 (Burton and Wells, 2002). The case of the Royal Liverpool Children's Hospital (Alder Hey) drew out parental feelings towards the remains of their children when it was discovered that pathologists had been 'systematically stripping dead children of their organs at autopsy and storing them, ostensibly for research purposes' (Carvel, 2002:55). Large collections of modern infant and child skeletal remains of known age, sex, ancestry and cause of death are rare, although some collections of paediatric skulls exist (Shapiro and Richtsmeier, 1997).

1.4.1   The child and the law

In 1998, the number of missing children in the UK was reported at 80 000 by the International Centre for Missing and Exploited Children (ICMEC). Cases of child murder in the USA have risen by 50% in the last 30 years and between 40 and 150 forensic cases involving children are handled annually (Morton and Lord, 2002). In England and Wales, children less than 1 year of age are most at risk of homicide (82 offences per million), compared to the overall risk of 15 per million in the total population (Intelligence and Security Council, 2000). Humanitarian investigations, such as those carried out by the Argentine Forensic Anthropology Team (EAAF) have recovered children's remains from mass graves in Guatemala and Argentina. An investigation of the Dos Rs massacre, in El Peten (1982), revealed that 47.3% of the listed victims were children, with the youngest victim only a few days old. During the excavation of a mass grave in Kibuye in Rwanda 202 (44%) of the 460 bodies recovered were of children (17% under 5 years; 14% 5–10 years; 13% 10–15 years) most showing evidence for blunt force trauma. However, the figures are considered to be an underestimate as decomposition and disarticulation of the remains meant that many smaller bones could have been missed (Schmitt, 1998). Identification of victims using DNA analysis is futile if no family members survive (EAAF, 1995). It is against the International Labour Organisation (ILO) Convention of 1982 for children under the age of 18 years to undergo compulsory recruitment in state armies. Nevertheless, voluntary recruitment is legal for those of 15 years and over under the Optional Protocol to the Convention on the Rights of the Child on the Involvement of Children in Armed Conflict, which came into force in 2002 (Harvey, 2003). Despite these rulings, in some conflicts child soldiers are as young as 8 (Dufka, 1999). The presence of children younger than 15 years in a mass grave argues against claims that these graves represent soldiers killed during legitimate conflict.

Another way in which children may enter the forensic record is as the result of suicide. Suicides of children and adolescents are rare, but when they occur, they generally involve 'fail-safe' methods such as hanging, running in front of a train or jumping from a height (Schmidt *et al.*, 2002). Motives include a break-up of a relationship, conflicts at school or in the home, mental illness or chronically disturbed family structures. Most at risk are usually males between 10 and 15 years, with cases rising into late adolescence and young adulthood. In Turkey, Ağritmiş and colleagues (2004) reported 43 cases of suicides in which 72% comprised females, perhaps as a result of the social status of females and their early marriage. Psychology studies have shown that children do not have a distinct perception of death until the age of 7 or 8 years, and do not develop sophisticated reasoning until 12–14 years; hence their contemplation of suicide is less likely (Schmidt *et al.*, 2002).

1.4.2   Mute witness: cases of child identification

Thankfully, forensic anthropology cases involving infants and children are rare in non-conflict areas. One of the earliest cases of child identification in England is a historical one. In 1965, the skeleton of Lady Anne Mowbray, daughter of the fifth Duke of Norfolk was disinterred during a cemetery clearance in Westminster. Anne died in 1481 at the age of 8 years and 11 months, and the remains were examined to see if her identity could be confirmed (Fig. 1.2). Warwick (1986) established the age, and attempted stature and sex estimations based on the femora. The dental remains were examined by Rushton (1965) who identified agenesis of the second permanent molars in the mandible and maxilla. Agenesis, but of different teeth, was also identified in the dentition of 12-year-old Edward V, Anne's blood relative and his brother, Anne's husband, 10-year-old Richard Duke of York. Molleson (1987) re-examined two sets of skeletal remains discovered in 1674 in the Tower of London that were thought to represent Edward and Richard, who, legend has it, were murdered in the Tower in 1484. Age and sex estimates were established using Hunt and Gleiser's method (1955) of correlating the stage of dental eruption with the fusion of the hand and wrist bones. The original examination of the Princes' remains in 1933 by Tanner and Wright (Bramwell and Byard, 1989) asserted that the remains were related not only due to dental agenesis, but because both of the skeletons had wormian bones of the same shape and in the same place on their lambdoidal sutures. Debates are still raging about the identity of these remains, and further forensic investigations using chemical analysis may help to finally solve the mystery once and for all.

Perhaps the first modern forensic case in the UK was published in 1958 when Imrie and Wyburn were presented with the fragmentary remains of an unknown child recovered from a hillside in England in 1955. With no previous cases to guide them, Imrie and Wyburn (1958) reconstructed the growth cartilage for the long bones from plasticine to estimate stature, and the triradiate cartilage of the pelvis to estimate sex, and managed to identify a likely individual. In 1967, the Oklahoma City abductions in the USA presented anthropologists with a particular problem. Although two girls aged 5 and 6 years had disappeared and were presumed dead, only one set of remains were recovered. The bones, a dress and hair were examined by Clyde Snow (Snow and Luke, 1970) who attempted to identify which child had been found and to rule out that the remains belonged to an entirely different child. After the Branch Davidian

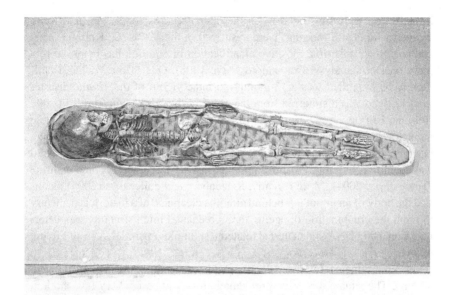

Figure 1.2 Remains of Anne Mowbray in her coffin. From Clark and White (1998:69), reproduced with kind permission from the Museum of London.

Compound incident near Waco in Texas in 1993, the bodies of 17 children were recovered from the ruins. Their preliminary identification was based on age assessments from the dentition and long bone measurements, in addition to their association with the adult remains (Owsley *et al.*, 1995). In cases of such commingling of non-adult bones, an assessment of the MNI requires specialist knowledge of their skeletal elements (Scheuer, 2002), for example being able to determine the side of a pars lateralis (part of the occipital) to determine duplication.

The recovery of Stephan Jennings was the first case in the UK to involve both an archaeologist and anthropologist in his recovery and identification. His remains were identified on the basis of his leather sandals and estimated age of around 3 years (Hunter, 1996). Careful examination by the anthropologist also revealed that the rib fractures were peri-mortem rather than post-mortem, and not caused when the stone wall that covered his body collapsed (C. Roberts, pers. comm.). One of the most infamous child murder cases in the UK was the Moors Murders. In 1965, the partially preserved bodies of Lesley Anne Downey and John Kilbride were recovered from the Moors and identified mainly on the basis of their clothing. The later recovery of Pauline Reade in 1986 followed the advent of DNA analysis; however, this was unsuccessful and again, her body was identified from her clothing. The problems of running a long-term investigation were illustrated when the dental records for Reade (and Keith Bennett who is still missing) were destroyed in the 1960s (G. Knupfer, pers. comm.). Since this time, positive identification of children has begun to rely more on DNA analysis (Yamamoto *et al.*, 1998). One of the youngest children to be identified was a 13-month-old male victim of the *Titanic* disaster in 1912, who was disinterred from the grave of the 'Unknown Child' in 2001 from the Fairview Lawn Cemetery in Halifax, Nova Scotia. Due to the acidic nature of the soil, only a few fragments of bone and three teeth were recovered, but mitochondrial DNA (mtDNA) was extracted from the teeth and the child was identified as Eino Viljami Panula. His grave has since become a shrine (Titley *et al.*, 2004). More recently, Katzenberg and colleagues (2005) identified the body of an infant left behind after the clearance of a historical cemetery through the combination of ageing through enamel microstructure, sex determination using DNA and maternal relatedness employing mtDNA from a living relative.

For adult remains, forensic odontology plays a leading role in confirming an identity. The situation is more problematic for child cases. Very few children are taken to see a dentist before the age of 5 years (O'Brien, 1994) and few have dental radiographs taken before the age of 10 for safety reasons. Dental radiographs are then kept for 11 years for adults, and for children they are held for 11 years or until they are 25 years old, whichever is longer, as regulated by the Consumer Protection Act 1987. Where dental charts do exist, the limited amount of dentistry experienced by young children means that there is a greater chance that their charts will be identical, with teeth at the same stage of development. Such was found to be the case at Aberfan (Knott, 1967) when in 1966, 116 children of the same age were killed when a waste tip slid down the mountainside into the mining village, engulfing their school. An alternative method for identification is smile comparisons. Using this technique, a photograph in life of the individual smiling is overlaid on a photograph of the deceased's front teeth. The morphology and orientation of the teeth from the image and deceased can then be directly compared (Perzigian and Jolly, 1984). With recent advances in technology, digital home videos can be used to provide comparisons for children where dental records do not exist (Marks *et al.*, 1997). Dental age estimated from a smiling photograph was instrumental in securing the conviction of one of England's most notorious serial killers. The case of Fred and Rosemary West began with the discovery of three bodies in the garden of 25 Cromwell Street, Gloucester in 1994 (Knight and Whittaker, 1997). Charmaine West, the daughter of Rosemary, was suspected to have been killed at their previous home at 25 Midland Road. The positive identification of one of the skeletons as Charmaine proved problematic when the same teeth, on opposite sides of the jaw, provided the forensic odontologist with two different ages, a year apart. When a photograph of the child was finally acquired, both a positive identification and the relative time of her death could be ascertained by comparing the extent of dental eruption in the skeleton to that of the photograph, which also had the date it was printed on the back (Knight and Whittaker, 1997). During the period in which the photograph was taken and Charmaine was killed, Fred had been in prison. Rosemary was convicted of murder.

1.5   Summary

The health and survival of the children from any given community is believed to represent the most demographically variable and sensitive index of biocultural change. Patterns of infant and child mortality have been shown to have a profound effect on the crude death rates of a population and, when coupled with evidence of childhood morbidity, have become accepted as a measure of population fitness. Roth (1992) described childhood as the most sensitive portion of the human life cycle, but it has taken us a long time to fully realise this potential. Only in the last decade have studies into the growth, development and health of non-adults been used to provide insights into the impact of the physical and cultural environment on the children themselves. The osteological study of non-adults occurred independently and at a much earlier period than theoretical concepts of childhood in social archaeology, but both have now reached a level of sophistication that should encourage communication and integration of the disciplines. Only then can we develop a fuller understanding of what it was like to be a child in the past.

2 *Fragile bones and shallow graves*

[t]he bone of infants being soft, they probably decay sooner, which might be
the cause so few were found here.

> *Thomas Jefferson (1788) commenting on remains from a Virginian*
> *earthen mound (cited in Milner et al. (2000:473))*

2.1 Introduction

One of the most commonly perceived limitations in the study of children from
archaeological contexts is their poor, or total lack of, preservation. The absence
of neonatal remains, or 'infant-sized pits' that no longer contain their 'tiny
occupants' (Watts, 1989:377), has led to the belief that children rarely survive
the burial environment, and even 'dissolve' in the ground. This issue has resulted
in the widespread neglect of studies into infant and child skeletal remains,
whose numbers are often believed to be too small for statistical analysis and
meaningful research. However, large numbers of non-adult remains have been
recovered from cemetery sites, and continue to be housed in museums and
universities ready for study (Table 2.1). In the UK, for example, the largest
include the multiperiod site of St Peter's Church, Barton-on-Humber ($n =$
c.1000), Romano-British Poundbury Camp ($n = 395$) and St Mary's Spital
in London, where so far, 1740 non-adults have been identified (White, pers.
comm.). In North America, sites such as Indian Knoll have yielded 420 non-
adults (Johnston, 1962). Collections of known age and sex also exist, most
notably the 66 non-adults from Coimbra in Portugal with a known cause of death
(Santos and Roberts, 2001). The oldest non-adult remains include a 2-year-old
Neanderthal child dating from over 100 000 years ago (Akazawa *et al.*, 1995).
In fact, at some sites non-adult remains can be better preserved than their adult
counterparts (Katzenberg, 2000). Buckberry (2000) reviewed the proportion
of non-adult remains recovered from Anglo-Saxon cemeteries and noted that
where adult bone preservation was good, so were the remains of the children.
In these cases, the percentage of non-adult remains recovered from the site
also increased. Just like adults, the histological structure of infant bones can be
excellent given the right burial circumstances (Colson *et al.*, 1997), and it has
been possible to extract bone proteins (Schmidt-Schultz and Schultz, 2003),
DNA and stable isotope signatures from their skeletons.

Table 2.1 *Archaeological sites with large numbers of non-adult skeletons*

Period	Site	Number of non-adults	Reference
Palaeolithic	Taforalt, Hungary	104	Acsádi and Nemeskéri (1970)
Neolithic	Isbister, Scotland, UK	156	Chesterman (1983)
Neolithic	Quaterness, Scotland, UK	72	Chesterman (1979)
Bronze Age	Lerna, Greece	140	Angel (1971)
Iron Age	Danebury, Hampshire, UK	50	Hooper (1991)
Pre-Dynastic	Naga-ed-Der, Egypt	123	Lythgoe (1965)
Dynastic	Gebelan and Asuit, Egypt	166	Masali and Chiarelli (1969)
4800–3700 BC	Carrier Mills, Illinois, USA	62	Bassett (1982)
3000 BC	Indian Knoll, Kentucky, USA	420	Johnston and Snow (1961)
150 BC–AD 750	Teotihuacan, Mexico	111	Storey (1986)
100 BC–AD 400	Gibson-Klunk, Illinois, USA	191	Buikstra (1976)
Roman	Intercisa and Brigetio, Hungary	107	Acsádi and Nemeskéri (1970)
Romano-British	Poundbury Camp, Dorset, UK	374	Farwell and Molleson (1993)
Romano-British	Lankhills, Hampshire, UK	118	Clarke (1977)
Romano-British	Winchester, Hampshire, UK	144	Roberts and Manchester (2005)
Romano-British	Cannington, Somerset, UK	148	Brothwell *et al.* (2000)
Anglo-Saxon	Raunds Furnells, Northamptonshire UK	208	Powell (1996)
Anglo-Saxon	Edix Hill, Cambridgeshire UK	48	Duhig (1998)
Anglo-Saxon	Empingham II, Rutland, UK	62	Timby (1996)
Anglo-Saxon	Great Chesterford, Essex, UK	83	Waldron (1988)
Anglo-Saxon	The Hirsel, Scotland, UK	153	Anderson (1989)
9th century AD	Sopronköhida, Hungary	72	Acsádi and Nemeskéri (1970)
AD 970–1200	K2 and Mapungubwe, South Africa	85	Henneberg and Steyn (1994)
AD 1050–1150	Meinarti, Sudan	101	Swedlund and Armelagos (1969, 1976)
AD 1300–1425	Arroyo Hondo, Mexico	67	Palkovich (1980)
13th century AD	Kane, Illinois, USA	60	Milner (1982)
Later Medieval	St Helen-on-the-Walls, York, UK	200	Lewis (2002c)
Later Medieval	Wharram Percy, Yorkshire, UK	303	Lewis (2002c)
Later Medieval	Westerhus, Sweden	225	Brothwell (1986–7)
Later Medieval	Jewbury, York, UK	150	Lilley *et al.* (1994)
Later Medieval	Fishergate, York, UK	90	Stroud and Kemp (1993)
Later Medieval	Royal Mint, London, UK	178	Waldron (1993)
Later Medieval	Trowbridge Castle, Wiltshire, UK	84	Graham and Davies (1993)
Later Medieval	St Nicholas Shambles, York, UK	51	White (1988)
Post-Medieval	Christ Church Spitalfields, London, UK	186	Lewis (2002c)
Post-Medieval	Bathford, Somerset, UK	63	Start and Kirk (1998)
AD 1500–1900	Ensay, Scotland, UK	200	Miles (1989)
AD 1750–1785	Larson, South Dakota, USA	441	Owsley and Bass (1979)
Multiperiod	St Mary Spital, London, UK	*c.*1740	W. White (pers. comm.)
Multiperiod	St Oswald's Priory, Gloucester, UK	128	Rogers (1999)
AD 1800–1832	Leavenworth, South Dakota, USA	211	Bass *et al.* (1971)

In 1973, Weiss estimated that 30–70% of a modern population would die before reaching 15 years of age, with the highest mortality levels in individuals between 1 and 5 years, and the lowest between 10 and 15 years of age (Weiss, 1973). Schofield and Wrigley (1979) reported that in modern Western populations the mortality rates of people under 10 years was only 2.4%, compared to 34% in pre-industrialised countries. It is now common to cite this modern pre-industrial figure as the norm for child mortality in archaeological populations from many periods all over the world. That 30% of the sample should contain non-adults has become the gold standard by which under-representation is measured. In many cemetery samples, the proportion of non-adults frequently fails to reach this target. Modern pre-industrial societies are not directly analogous with those in the past, and they are not untouched by the pressures of industrialisation that surrounds them. Factors that will affect infant and child survival, such as their weaning diet, exposure to pollution and disease, living conditions and economic status, are often dictated by the wealthier industrial countries. Therefore, it is open to debate whether 30% child mortality is the correct minimum we should expect from an archaeological sample where all the children are included.

Some authors have argued that the usually low numbers of infant remains in our archaeological populations may actually reflect true mortality rates (Brothwell, 1986–7; Panhuysen, 1999), but cemetery samples revealing only 12 individuals dying under 1 year of age, over a 500-year period, suggest that significant under-representation does exist (Lewis, 2002c). In the rare cases where burial records from an excavated sample survive, the proportion of infants reported to have died and those excavated are usually at odds. For example, the low number of infants from the Voegtly cemetery in Pittsburgh, where 311 infants were reported to have died but only 200 children under 5 years were recovered, was suggested by Jones and Ubelaker (2001) to demonstrated taphonomic bias in preservation. Conversely, Herring and co-workers (1994) found a greater number of infants in the skeletal sample from Belleview, Ontario than were listed in the burial records, perhaps as the result of poor record-keeping and the surreptitious burial of illegitimate and unbaptised babies. Nevertheless, the proportion of non-adults recovered from a cemetery is an important tool when trying to understand the nature of funerary treatment of the youngest members of society. Sellevold (1997) suggests that in the medieval period, different types of church would attract a different congregation, and this may be reflected in the age and sex distribution of those buried at the site. For example, she states that 15% of those buried at Hamar Cathedral in Norway were infants and children, compared to 22% in the small parish church of Sola. In the UK, the early medieval pre-Christian cemeteries consistently show a paucity of young children. The early Anglo-Saxon cemetery at Sewerby, North Yorkshire produced only 5% of children under 5 years old, with West Heslerton, also North Yorkshire providing only 6% (Lucy, 1994). In contrast, the earlier Roman cemetery of Owlesbury, Hampshire had 35% of children under 5 years, the later Anglo-Saxon cemetery at Norwich produced 45% child burials, and 48% of the individuals recovered from later medieval Winchester were children (Daniell, 1997:124). Lucy (1994) argues that this suggests that in the early Anglo-Saxon period children were buried elsewhere, and that poor preservation cannot account for the lack of infant graves in these sites when compared to earlier and later cemeteries. In such cases, the under-representation of non-adults is usually explained in terms of taphonomic, methodological and cultural factors.

2.2   Fragile bones?

Taphonomy is the scientific study of the processes of burial and decomposition that influence the preservation of remains, from the cause and manner of death, post-mortem anthropogenic alterations and animal scavenging, to the chemical and physical breakdown of tissues (i.e. diagenesis). Much of the work carried out into soft-tissue decomposition, and the influence of the burial environment on the survival of bone elements, has been carried out in North America on modern remains. These studies have shown that a complex interaction exists between many internal and external factors relating to the body and burial environment. These will affect the rate and nature of soft- and hard-tissue decay (e.g. age and sex, presence of disease, open wounds, soil pH, temperature and burial depth). As each case is unique, it is impossible to make broad assumptions about the sequence of events during body decomposition. Taphonomic processes on non-adult remains have not been fully documented in forensic archaeology due to the rarity of cases. However, a child's body can become skeletonised in just 6 days compared to several weeks for adult remains. Children's bodies become disarticulated more easily and are more susceptible to scavenging and dispersal as their small size makes it easier for animals to move them around (Morton and Lord, 2002). Their crania are also more fragile and prone to post-mortem damage (Haglund, 1997). As for burned remains, Holck (2001) has recorded that fetal remains can be completely destroyed in a house fire in under 20 minutes.

In forensic anthropology, it is often important to understand the rate of soft-tissue decay in order to estimate time since death, and to piece together a series of decomposition events that may have influenced the pattern and distribution of scattered remains. The soft-tissue composition of children's bodies means that they will decompose at a different rate to adults. Their bodies cool more rapidly due to a greater surface-area-to-weight ratio (Smith, 1955:18), and infants who die before their first feed will decompose at a slower rate than a fed infant, due to the lack of organisms in the stomach, intestines and lungs (Bell *et al.*, 1996). Human infants are born with both brown and white adipose tissue, which serves to insulate and store energy during growth (Kuzawa, 1998). This tissue composition means children are prone to mummification rather than saponification (adipocere formation). All these factors can have an affect on the time since death that may be estimated for a child, especially if the rate of decomposition is derived using animal analogues with lower levels of brown adipose tissue.

An unusual feature in non-adult remains is the purple–blue coloration often observed in the deciduous teeth and developing permanent crowns post-mortem. Pigmented teeth are associated with certain pathological conditions (e.g. fluorosis and congenital heart disease), and so-called 'pink teeth' in forensic contexts are indicative of trauma and asphyxiation, as blood is forced into the pulp cavities with the subsequent build-up of iron (Dye *et al.*, 1995). Detailed examination of these pigmented teeth from burial environments has so far shown the cause to be diagenetic. Archaeological 'pink teeth' are a consequence of fungal infiltration, and the purple–blue coloured teeth the result of porous dental enamel and the uptake of iron, manganese and organic matter solutions from the surrounding soil (Dye *et al.*, 1995; Mansilla *et al.*, 2003).

Once the body has been reduced to hard tissues, the pace of deterioration slows with bones often maintaining their external appearance for many hundreds of years, until eventually bacterial and plant activity work to undermine their chemical and physical structure. Living bone is made up of water, an inorganic or mineral component in the form of hydroxyapatite (calcium, phosphate and carbonate), and an organic matrix (principally collagen). The proportion of these elements is governed by the age of the individual and type of bone (i.e. compact or cancellous) being considered (Nielsen-Marsh *et al.*, 2000). As with soft-tissue decomposition, no single factor determines the level of bone preservation, and a complex relationship exists between a number of overlapping variables, both intrinsic and extrinsic. Intrinsic factors include the chemistry, size, shape, density, porosity and age of the bone, whereas extrinsic factors such as groundwater chemistry, clothing, soil type, temperature, oxygen levels, flora and fauna all play a role in diagenesis. In general, however, the decay of bone follows three phases: the chemical deterioration of the organic component, the chemical deterioration of the mineral component, and the microbiological attack of the composite form of the bone (Collins *et al.*, 2002).

During the decay process, the organic matrix is subject to slow hydrolysis of proteins to peptides and eventually amino acids, while the inorganic component undergoes spontaneous rearrangement of the crystalline matrix, weakening the protein–mineral content and leaving the bone susceptible to dissolution (Henderson, 1984). Microbial attack also has a profound influence on both the macroscopic and microscopic preservation of bone, increasing its porosity and hence its susceptibility to groundwater dissolution (Hedges, 2002). Although studies focussing on the differential rate of adult and non-adult bone decomposition are rare, Robinson and colleagues (2003) have shown that immature faunal remains are more porous than their adult counterparts, and that this factor makes them more vulnerable to decay. The high organic and low inorganic content of children's bones (Guy *et al.*, 1997), in addition to their small size, certainly means they may be more susceptible to taphonomic processes.

Gordon and Buikstra (1981:589) suggest that in high alkaline or acidic soils '. . . all or most of the infants and children may be systematically eliminated from the mortuary sample by preservational bias'. Walker and colleagues (1988) demonstrated this problem when examining the burial records of adults and children against skeletal remains recovered from Mission La Purisima, California. The burial records indicated that 32% (150 out of $n = 470$) of the people buried within the cemetery were under 18 years old, when only 6% (28) of the skeletons represented individuals of this age. Poor preservation of the remains as a whole was attributed to the acidic sandy soil in which they were buried, which allowed water to permeate through the bone, with subsequent soaking and drying disintegrating the fragile ribs and spine. The less dense cancellous bone (e.g. ends of the long bones, spine), covered with only a thin layer of cortical bone, has a larger surface area allowing for the more rapid decay of the collagen and mineral matrix as it is exposed to the groundwater (Von Endt and Ortner, 1984; Nawrocki, 1995). While cancellous bone decays more rapidly than cortical bone, the infant skeleton actually contains less trabecular bone than adults and older children, with the radius being the densest bone (Trotter, 1971). Bone mineral content also varies during childhood, with the lowest levels in the post-neonatal period, returning to the denser neonatal levels at the end of the second year. Thereafter, the mineral level steadily increases until the adult level is reached.

Ingvarsson-Sundström (2003) when analysing her non-adult samples from Greece, found that the small compact bones of the neural arches, phalanges, zygomatics and temporal bones survived and were recovered well, but that only 28.6% of the long bones were preserved well enough to be measured for growth and age. In the Inuit sites of Hudson Bay (AD 1200), Merbs (1997) noted small empty graves and that cranial and leg bones were best represented among the infants. In older children the scapulae, arms and pelvic bones survived best. Adolescents and adults showed a better, similar pattern of preservation. Therefore, it is expected that the recovery of non-adult remains will vary depending on the age group and burial environment, with the epiphyses being particularly susceptible to groundwater penetration, and fetal remains surviving well in most burial conditions.

2.3   Little people . . . little things . . .

The size and number of skeletal elements that make up the non-adult skeleton varies with age, but ranges from 156 recognisable bone elements at birth (Table 2.2), to 332 at around 6 years as the epiphyses appear and fuse (Table 2.3), compared to 206 bones in the adult skeleton. The number and complex morphology of the non-adult skeleton can make identification challenging. While excellent texts relating to non-adult skeletal anatomy exist (Fazekas and Kósa, 1978; Scheuer and Black, 2000, 2004; Baker *et al.*, 2005), the recovery of the small bones, epiphyses and developing teeth can be hindered by lack of expertise of the excavator, and will have a great impact on the information that can be gained from the remains. Due to their porosity, immature remains are more susceptible to colour change in the soil, making them harder to recognise, and the epiphyses may mimic small stones and can easily be missed by the untrained eye. The burial matrix, such as sandy and clay soil or gravel and stone inclusions, can also severely hinder the identification and successful recovery of infant bones. Rarely are diagrams outlining the bones of the non-adult skeleton provided on a field recording form, and these at least would aid the excavator in ensuring that all the bony elements are accounted for (Fig. 2.1).

Dental crowns develop in large alveolar crypts in the mandible and maxilla. Once the gum (gingiva) has decomposed, this often results in the loss of fragile and tiny developing crowns during excavation, processing and storage. The pre-wrapping of all fetal and infant mandibles in tissue paper directly after excavation is recommended to help preserve these tooth germs (Spence, 1986, cited in Saunders, 2000). Another common problem with tiny remains is that they are often incorrectly identified as non-human, particularly as bird or rodent bones. Several researchers have recounted tales of early excavations where fetal remains were found with animal bone material that the excavators failed to recognise (Robbins, 1977; Wicker, 1998; Ingvarsson-Sundström, 2003). The late discovery of the cremated remains of a 34–36-week fetus, originally thought to be animal bone, among those of the 'rich Athenian lady' at Athens changed the interpretation of the burial tomb. The ritualised burial and wealthy gravegoods were reconsidered as reflecting the fact that she died young during pregnancy or childbirth, rather than purely being a statement of her economic status (Liston and Papadopoulos, 2004). For the long bones at least, the fusion of epiphyses, the density and lighter staining of the cortex in animal remains should help distinguish them from human. It is important

Table 2.2 *Number and type of bone elements expected in a neonate*

Element	Specific element	Number	Element	Specific element	Number
Frontal		2	Sacrum	Centrum	5
Parietal		2		Lateral elements	6
Occipital	Pars basilaris	1		Neural arch	10
	Pars lateralis	2			
	Pars squama	1	Manubrium		1
Temporal	Pars squama	2	Sternum		4
	Petrosa	2	Ribs		24
	Tympanic ring	2 (attached)	Clavicle		2
Maxilla		2	Scapula		2
Zygomatic		2	Humerus		2
Mandible		2	Radius		2
Sphenoid	Greater wings	2	Ulna		2
	Body	1	Carpals	Capitate, hamate	4[a]
Nasal		2	Metacarpals		10
Ethmoid		2	Phalanges	Proximal	10
Concha		2		Intermediate	8
Vomer		1		Distal	10
Lacrimal		2	Pelvis	Ilium	2
Hyoid		3[a]		Pubis	2
Atlas	Neural arch	2		Ischium	2
Axis	Dens	1	Femur		2
	Centrum	1	Tibia		2
	Neural arch	2	Fibula		2
C3–C7	Centrum	5	Tarsals	Talus	2
	Neural arch	10		Calcaneus	2
T1–T12	Centrum	12	Metatarsals		10
	Neural arch	24	Phalanges	Proximal	10
L1–L5	Centrum	5		Intermediate	8
	Neural arch	10		Distal	10
Total					263

[a]Elements may not always appear during this period.

to remember that fetal and juvenile animal bones with unfused epiphyses are also recovered, and a working knowledge of non-adult skeletal morphology is crucial.

It is a general rule that the deeper the grave the better preserved the body as it is protected from fluctuations in temperature, scavengers and removal from the original site of deposition (Henderson, 1984). Scull (1997) noted that the infant and child burials at Watchfield cemetery in Oxfordshire were in shallow

Table 2.3 *Number and type of bone elements expected in a 6-year-old child*

Element	Specific element	Number	Element	Specific element	Number
Frontal		1	Clavicle	Shaft	2
Parietal		2	Scapula	Body	2
Occipital	Pars squama	1		Coracoid epiphysis	2
	Pars basilaris	1[a]	Humerus	Shaft	2
Temporal		2		Proximal epiphysis	2
Maxilla		1	Radius	Shaft	2
Zygomatic		2		Proximal epiphysis	2
Mandible		1	Ulna	Shaft	2
Sphenoid		1		Distal epiphysis	2
Nasal		2	Carpals	All but pisiform	14
Ethmoid		1	Metacarpals	Shafts (epiphyses)	10 (10)
Concha		2	Phalanges	Shafts (epiphyses)	28 (36)
Vomer		1	Pelvis	Ilium, pubis, ischium	6
Lacrimal		2	Femur	Shaft	2
Hyoid		1		Proximal epiphysis	2
Atlas	Neural arch	1		Distal epiphysis	2
	Anterior arch	1[a]	Tibia	Shaft	2
Axis		1		Proximal epiphysis	2
C3–C7		5		Distal epiphysis	2
T1–T12		12	Patella		2
L1–L5		5	Fibula	Shaft	2
Sacrum		5		Proximal epiphysis	2
Coccyx		2		Distal epiphysis	2
Manubrium		1			
Sternum		5	Tarsals		14
Xiphoid		1[a]	Metatarsals	Shafts (epiphyses)	10 (18)
Ribs		24	Phalanges	Shafts (epiphyses)	28 (36)
Total					232

[a]Elements may not always appear during this period.

graves. The infant burials that were recovered were found within or at the base of ploughsoil, and therefore were more susceptible to ploughing or machine stripping before excavation. Ingvarsson-Sundström (2003) reported that the most frequent type of grave for children in the Lower Town of Asine was pit graves, which were overlooked in favour of the more obvious built graves in earlier excavations. Although depths of children's graves from archaeological investigations are rarely reported, Murail and Girard (2000) showed that children under 15 years of age were buried at 1.40 m compared to 1.56 m for adults and older children in the fifth-century rural cemetery of Chantambre, France. Yet, in a previous study of two medieval cemeteries in Maastricht, The Netherlands, Panhuysen (1999) found no difference in the depths of graves

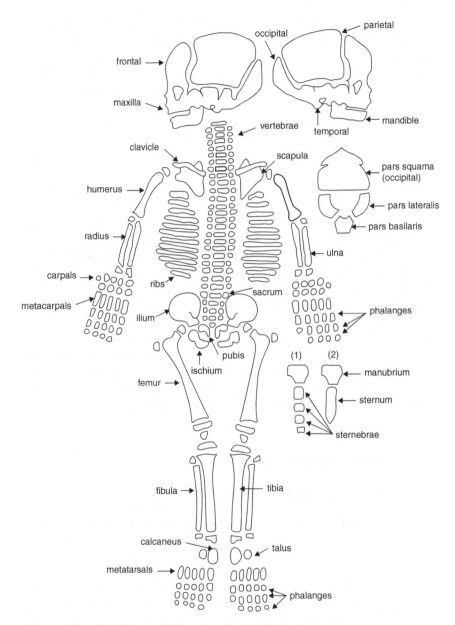

Figure 2.1 Diagrammatic inventory of non-adult bones.

between adults and non-adults. Sellevold (1997) noted that a larger number of dispersed bones from Sola and Simonsborg in medieval Norway were of non-adults, and that their graves were more prone to disturbance perhaps due to their shallow nature. Crawford (1991) has suggested that shallow burials of children indicate the low regard in which they were held; however, she later reasoned that it may be more due to practicalities and less of a need to bury at depth, reflecting that it is harder to dig a deep grave of small dimensions (Crawford, 1993). In fact, Sellevold (1997) noted that grave length was not related to adult or non-adult burials and that there was great variation in the size of non-adult graves from the age of 1 year.

Once excavated, the importance of the analysis of non-adult remains is not always recognised. Robbins (1977:15) reported that, along with fragmentary remains, infant bones were 'not saved for analysis' from the Carlson Annis shellmound in Kentucky, USA. In Scandinavia, before the 1980s, it was the practice of museums to preserve only the intact skulls and, as children's skulls are made up of several fragile bones, they were generally disposed of (Sellevold, 1997). Buckberry (2005) reported that nearly half of all the infant remains from St Peter's Church, Barton-on-Humber, Lincolnshire were located in the disarticulated material, as were the majority of infants from Fillingham, Lincolnshire. These remains were probably originally in shallow graves and disturbed by subsequent graves and ploughing. As many post-excavation examinations do not include disarticulated or unstratified bones due to lack of time and money, numerous child skeletons are probably hidden and will remain unstudied in the stores of archaeological units and museums.

2.4   The marginalised child?

> If any child be deadborn [she was to] see it buried in such a secret place as neither hog nor dog nor any other beast may come onto it [nor] suffer any such child to be cast into the jaques [latrines] or any other inconvenient place.
> Oath of eighteenth-century midwives (Lane, 2001:121)

The earliest evidence for the special treatment of children in death comes from the extraordinary burials from the Upper Palaeolithic. Formicola and Buzhiova (2004) reported on the double burial of two children aged 9–10 and 12–13 years old from Sunghir in Russia. The remains were found head to head, with ivory spears and rich gravegoods, and covered in red ochre. Palaeopathological examination of the remains suggests that both were suffering from congenital abnormalities producing bowing of the femora. This burial adds weight to the theory that in the Upper Palaeolithic, visibly disabled individuals were afforded special burial rites, first implied by the excavation of the triple burial of the same date from Dolní Věstonice (Trinkaus *et al.*, 2001). Red ochre appears again in a Neolithic collective burial from Ertebølle in Zealand, The Netherlands, where several adult males and females were buried with infants and children. One male, the last burial to be placed in the grave, was described as 'cradling' the body of an infant in his arms. All the remains were sprinkled with red ochre 'seemingly most thickly around the children' (Whittle, 1996:198). Boric (2002) discusses the numerous red-ochre-covered infant burials from Vlasac and Lepenski Vir in Neolithic Serbia, and in particular, the burial of a female with red ochre covering her pelvis which contained a fetus. The significance of red ochre is thought to symbolise blood, female menstruation, ancestral blood ties and bloodshed. Boric (2002) also argues that the smothering of infant burials with red ochre from Lepenski Vir represents protection. Perhaps not yet considered members of that society, infants would not have undergone the protective rites of passage afforded to the adults and older children. The discovery of a late Iron Age burial of a 2-year-old from Bath in England provides more evidence that in some periods at least, children were afforded exceptional burial rites, perhaps due to illness. The child was found under a limestone slab, next to an adult male with signs of tuberculosis. The child had a systemic inflammatory condition that would have been visible during life. Both burials were inhumations in a cemetery full of cremated individuals, and both were situated in an area considered to be a sanctuary (Lewis and Gowland, 2005).

Examples of children's remains located outside the confines of communal burial grounds are a common finding throughout the world, and during all time periods. Infant remains have been found under the floors of the world's earliest agricultural settlements in Natufian sites in the Near East (12 500–10 200 BP) (Scott, 1999), and their remains have been found within domestic spaces from the Neolithic to the Roman period. These burials have been variously interpreted as the result of infanticide, as symbolising domestic or gendered space (Scott, 1990), as due to a lack of social ranking, as a result of economic necessity or as sacrificial burials (Watts, 1989; Wicker, 1998). These marginal burials were not universal. For example, carefully wrapped fetal remains have been recovered within the communal cemetery of the Roman Dakhleh Oasis in Egypt (Tocheri *et al.*, 2001). In the Mesolithic, infant burials have been located within communal cemeteries, with children's graves at Vedbaek, Denmark associated with red ochre and animal bones, and one infant in particular being laid under a swan's wing, with a broken flint blade (Thorpe, 1996). The small number of child burials in many of these periods suggests that the vast majority may have been deposited elsewhere (Scott, 1999). The distinct absence of children under 1 year of age at Mokrin in former Yugoslavia indicates that they were purposefully excluded and not yet considered part of the community

(Rega, 1997). Child cemeteries have been identified at Yasmina in Carthage (present-day Tunisia) (Norman, 2002) and Lugano in Italy (Soren and Soren, 1999) and in the historic periods of Ireland, children were buried in separate and marginal cemeteries or 'cílliní' (Finlay, 2000). Evidence for one such burial ground has also been reported in Wales (Manwaring, 2002). In some communities there is evidence that children underwent a different type of burial. For instance, Chamberlain (1997) reported a higher number of child burials in caves (39%) compared to long barrows (25%) in the British Neolithic, a pattern also reflected in southern Europe. Cauwe (2001) found children's remains clustered in the northern sector of a Mesolithic tomb in Abri des Autours, southern Belgium, an open area without natural protection, whereas the majority of the adult burials were located in the southern sector in a more protected area. There was also some evidence for the deliberate removal of their skulls.

In the later Anglo-Saxon and medieval periods, when children are included in the communal cemeteries, they are often found to cluster, but these clusters are not universally defined or easy to predict. For example, at St Andrews Fishergate in York, 76% of the under-5s were found in the western third of the cemetery (Stroud and Kemp, 1993), while excavations at Raunds Furnells (Northamptonshire), Jarrow (Tyneside), Whithorn Priory (Galloway, Scotland) and Winchester (Hampshire) have revealed groups of infants along the east side of the church, where perhaps it was thought 'holy' water dripping off the eaves would baptise them (Crawford, 1993). This practice seemed to die out after 1066, and later child burials usually appeared in either the west or eastern ends of the cemetery (Daniell, 1997). In St Olav's monastery in Trondheim, Norway infants had a burial place reserved for them close to the choir, which if left unexcavated, would have meant the loss of 45% of the infant remains. At Hamar Cathedral, Norway 57% of the remains recovered from the south-west of the cemetery were children compared to only 8% in the west (Sellevold, 1997). Similarly, at Taunton (Somerset) in England, 85% of the non-adults were located at the western end of the cemetery, while only three children were excavated from the east (Rogers, 1984).

At Deir el Medina in Egypt, the cemetery seemed to have been divided into age-determined sections, with the lowest part of the slope reserved for infants and perinates, and even placentas in bloody cloths (Bruyère, 1937, cited in Meskell, 1994). The infants were buried in containers associated with the household including amphorae, fish baskets, boxes and chests, with some small children and adolescents buried in roughly made coffins (Meskell, 1994). Ethnographic sources indicate that prehistoric Iroquoians of Southern Ontario buried infants along pathways believing their souls would influence the fertility of passing women. Excavations of Iroquoian village sites have been interpreted as reflecting this belief, with large numbers of infants located under the central corridor of longhouses, perhaps acting as a 'path' on which infertile women walked (Saunders, 2000). If the whole area of the cemetery is not excavated, as is often the case in urban developments, this could lead to an under-representation of the youngest members of the society. At St Helen-on-the-Walls in York, UK the recovery of only 12 infants was considered the result of infant burials being clustered at the north side which remained unexcavated (Dawes and Magilton, 1980). However, this hypothesis later proved ill-founded, and the infants appear to have been afforded another, unknown location for their burial (Grauer, 1991).

In the later medieval period, Protestant and Catholic dogma influenced the inclusion of infants within cemeteries. Newborns were considered to be corrupted by the original sin of their conception, and unbaptised or stillborn infants were not permitted burial in consecrated ground (Orme, 2001:24). However, in 1398 a royal licence was given to enclose the Hereford Cathedral cemetery to prevent 'the secret burials of unbaptised infants' (Daniell, 1997:127). This practice seems to be substantiated by the discovery of 24 infants shallowly buried in a haphazard manner in an extended part of the cathedral cemetery (Shoesmith, 1980). Daniell (1997:128) suggests this is the nearest equivalent to a child cemetery that exists in medieval England.

In seventeenth-century England, there is documentary evidence for the disposal of illegitimate children or infanticide victims in gardens, bogs, under the floorboards or hidden in boxes, with some women entering the churchyard at night to bury their babies (Gowing, 1997). We know little of the fate of the hundreds of infants who died in 'baby farms' (where abandoned infants were taken in by the parish), or whose bodies were discovered in the streets. Who claimed their bodies from the Coroner? And if they were afforded burials, where were they buried? In 1862, 13 unregistered infants were found within St Peter's Churchyard in London, where the sexton was running a lucrative business charging a shilling for the clandestine burial of 'stillborn' infants (Behlmer, 1979). Moreover, in the eighteenth and early nineteenth centuries, the practice of using wet-nurses meant that children from towns and cities who died whilst being nursed in the countryside were often buried there rather than being returned home (Fildes, 1988a). This type of infant migration should be taken into account when attempting to assess infant mortality levels in urban and rural samples of this date.

2.5   Obstetric deaths

Infant remains may be better recovered when they are present in deeper, more distinct adult graves, in particular, with females who died before or during childbirth. These tiny bones may be overlooked in hurried excavation when sieving is not possible. Today, obstetric death rates for women between the ages of 15 and 44 years in the UK are reported at 58.4 per 100 000, with deaths during pregnancy low when compared to those of women 43 days post-partum (Ronsmans *et al.*, 2002). Obstetric deaths can result from puerperal fever, dystocic presentation, obstructed labour due to a malformed pelvis, breach birth or an oversized fetus (e.g. as the result of diabetes), haemorrhage or premature detachment of the placenta. In the past, such hazards and infections would have been similar. Although deaths during delivery or due to a lack of prenatal care, instruments or appropriate and qualified assistance may have been higher in the past. In 1775, the London Bills of Mortality reported 188 deaths of women in 'childbed' (Roberts and Cox, 2003:316) but this would certainly have been an underestimate, if only a small one, as deaths of women with a child still inside the womb were omitted from the records (Eccles, 1977).

   The presence of female burials with fetal remains in the archaeological record can provide evidence for these hidden deaths. At least 24 cases have been reported from excavations around Britain (Roberts and Cox, 2003), with further isolated cases reported throughout the world. Arriaza and colleagues (1988) reported 18 cases (14%) of childbirth related deaths in Chilean mummies dating from 1300 BC to AD 1400, with one woman showing a full-term fetus in breach position, and another described, dramatically, as clutching her umbilical cord as she struggled to deliver a second child. When soft tissue is not preserved, recognition of such deaths can only be made through meticulous excavation, where an intimate link between the female and infant burial needs to be demonstrated. The position of the child's head or leg bones directly within the pelvic cavity (i.e. overlying the sacrum between the ischium and pubic symphyses), would be evidence of this link. Some of the earliest potential evidence for death during childbirth was presented by Smith and Wood-Jones (1910), who reported two cases of Egyptian females with deformed pelvises, perhaps as the result of rickets, and fetal heads in the pelvic cavities.

   The careful excavation of such neonatal remains is essential to establish whether the child was placed within the grave after birth, was contained within the pelvic cavity (in utero) or had been expelled during decomposition (Smith, 1955:25). For example, at Castledyke, Barton-on-Humber, a female carrying a 37-week fetus appears to have been placed with her right hand over her extended abdomen (Fig. 2.2), as hand and fetal bones were found to the left of the lumbar spine, and the skull of the fetus rested on the left ilium in the birth position (Boylston *et al.*, 1998). Post-mortem disturbance of a double grave at Whitehawk Neolithic camp in Brighton (Sussex, UK) made it unclear whether the infant recovered was associated with the female (Curwen, 1934). Although fragments of the child's skull were found adhering to the inside of the female pelvis, the age of the infant suggested the child had been placed in the grave

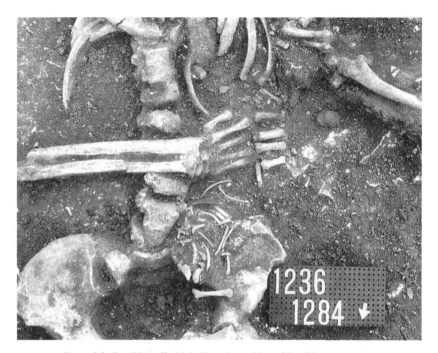

Figure 2.2 Possible coffin birth. Note the position of the tibia under the pubic symphyses. From Chadwick Hawkes and Wells (1975:49), reproduced with kind permission from Barry Cunliffe.

after birth. Ancient DNA analysis could be used to assess the relationship of the mother to the child. This distinction is essential as the placement of a child within an adult grave (not necessarily that of a parent) as the result of coinciding deaths was part of the Anglican tradition (Roberts and Cox, 2003:255). A child may have been placed on top of the chest or stomach, resulting in bones coming to rest within the pelvic cavity during decomposition. Two possible examples of this practice come from St Helen-on-the-Walls in York, where a 9-month-old child lies outside the abdominal cavity, on the left side of the pelvis, in one case; and in another, a woman appears to have had a child laid to rest on her stomach (Dawes and Magilton, 1980). Gilchrist and Slone (2005:71) report on the guidelines of an Augustinian canon, John Mirk, dating to the late fifteenth century, which states that: 'A woman that dies in childing shall not be buried in church, but in the churchyard, so that child should first be taken out of her and buried outwith the churchyard.' Evidence from monastic cemeteries indicates that the notion of impurity of parous women led to some children being left within the womb. Tiny infant burials placed within the mother's grave, for example in a small coffin as at Hull Augustinian Friary (Gilchrist and Sloane,

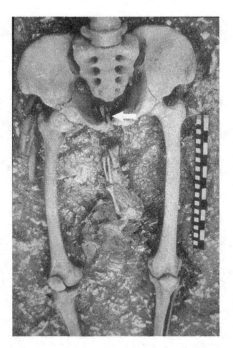

Figure 2.3 Female with fetal remains on the left side. It is not clear if the child was placed on top of the female or comes from within the womb. From Drinkhall and Foreman (1998:366), reproduced with kind permission from The Continuum International Publishing Group.

2005), may have been the result of obstetric hazards after all, but the babies were removed from the womb post-mortem.

Møller-Christensen (1982) described a so-called 'coffin birth' in the Aebelholt monastery in medieval Denmark. A coffin birth is supposed to occur when the decomposition of the womb results in the post-mortem expulsion of the unborn fetus, usually 48–72 hours after the death of the mother, but this phenomenon has yet to be reported in the clinical or forensic literature. This event was suggested at Kingsworthy Park (Hampshire, UK) (Chadwick Hawkes and Wells, 1975), where the infant's feet were still located under the pubic symphyses (Fig. 2.3). Two Arikara females with fetuses apparently in utero were recovered from the Larson site, South Dakota, USA (Owsley and Bradtmiller, 1983), and an additional female was buried with 'twins' aged 24 weeks. Malgosa and colleagues (2004) presented a case from Bronze Age Spain (1500–1000 BC) of a female with a child in an abnormal position during birth, with the right arm of the child extended and located under the pubic symphyses, perhaps as the result of a transverse lie birth presentation. The authors suggest that death was due to haemorrhage and exhaustion on the part of the mother. At St Nicholas Shambles in London, a fetus was found face down and 'wedged' into the pelvis (Wells, 1978), and at St Clements, Visby in Gotland, Sweden, the fetus and young adult female both showed multiple exostoses that, it is argued, obstructed the birth canal and led to their deaths (Sjøvold *et al.*, 1974). Wells (1975a) has stated that a deformed pelvis was the only direct evidence of an obstetric death in the archaeological record.

The discovery of a female with fetal remains has important implications for the analysis of trace elements, where pregnant and lactating women are reported to have lower concentrations of zinc and calcium, but a higher concentration of strontium (Blakely, 1989; King, 2001; Markland, 2003), although diagenetic factors are a cause for concern. Studies of nitrogen levels in the hair of pregnant females also suggests a decrease in $\delta^{15}N$ values at birth compared to maternal pre-pregnancy values (Fuller *et al.*, 2004), which may be reflected in archaeological human bone collagen. Extrapolating such evidence could allow us to analyse the number of lactating females within a cemetery, refine our estimate of maternal mortality, and aid in the identification of pregnant and lactating women within the forensic context.

2.6   Summary

Arguments that the remains of infants and children will not survive the burial environment are unfounded, and in large cemetery excavations, non-adult remains can be found in sufficient numbers to allow in-depth research into their morbidity and mortality to be carried out. Evidence from the taphonomic literature suggests that, while children's bodies may decompose more quickly, their bones have the potential to survive well in the same conditions that allow for the good preservation of adult skeletal remains. The paucity of non-adults in a cemetery sample can be the result of many additional factors, including the skill of the excavator, curatorial bias and cultural exclusion. Funerary treatment of non-adults in the past does not follow any specific pattern and varies in extremes from the carefully placed ritualistic graves of the Upper Palaeolithic and the child cemeteries of Hellenistic Greece and Rome, to their unceremonious burial in rubbish pits in post-medieval England. Patterning of graves within communal cemeteries, where children are clustered in specific sections of the graveyard for practical or ritual reasons, will also limit their recovery if the entire cemetery is not excavated.

3 *Age, sex and ancestry*

3.1   Non-adult ageing

In bioarchaeology, the age-at-death of a child is used to make inferences about mortality rates, growth and development, morbidity, weaning ages, congenital and environmental conditions and infanticide. In forensic contexts, assigning an age to a living child of unknown identity may be necessary when the child is suspected of a crime; when penal codes differentiate law and punishment for children of different ages (Schmeling *et al.*, 2001; Foti *et al.*, 2003); or if the child is a refugee of uncertain age. For the deceased child, age estimation is considered to be the most accurate biological identifier that a forensic anthropologist can provide.

Age estimation of non-adults is based on a physiological assessment of dental or skeletal maturation, and relies on the accurate conversion of biological into chronological age. Error in the accuracy of this conversion can be introduced by random individual variation, the effects of the environment, disease, secular changes and genetics (Demirjian, 1990; Saunders *et al.*, 1993a). Most importantly, the age of development of the dentition and skeleton are known to differ between the sexes, a biological assessment that has yet to be carried out successfully in non-adults.

3.1.1   *Dental development*

Dental development (mineralisation and eruption) is less affected by environmental influences than skeletal growth and maturation (Acheson, 1959), and mineralisation of the dentition is the preferred method for producing an age estimate for non-adults. The deciduous teeth begin to mineralise in the jaw around 15 weeks gestation, beginning with the maxillary central incisors and continuing until all of the deciduous teeth are fully erupted, when the child is around 3 years of age. Development of the permanent dentition covers a period from birth to around 14 years of age, with the eruption time of the third molar considered the most variable, but roughly occurring around the age of 17 years (Scheuer and Black, 2000). Differences between the sexes are evident, with females being on average 1–6 months ahead in their overall dental development than males. This variability is also evident in individual teeth, with the canines considered the most sexually dimorphic, and females being as much as 11 months ahead of males in their development (Demirjian and Levesque, 1980).

There is a wealth of literature on the development and eruption of the deciduous and permanent dentition in children. Standards exist for various modern and ancient populations (Table 3.1) which employ both radiographic and macroscopic techniques. Dental development follows a typical pattern and can be divided into convenient stages, beginning with cusp mineralisation of the deciduous maxillary incisors, and normally ending in root apex closure of the third molar. The timing and sequence of this development differs between populations (Lovey, 1983; Harris and McKee, 1990; Tomkins, 1996; Holman and Jones, 1998; Willems *et al.*, 2001) and the calcification of the third molar in particular has been shown to be highly variable (Fanning and Moorrees, 1969; Kullman, 1995; Bolaños *et al.*, 2003). Such discrepancies can cause problems when trying to assign a dental age to an individual with a different sequence pattern to the standard being used, resulting in a mixture of younger and older ages based on different teeth within one jaw (Owsley and Jantz, 1983). This variation has also been noted in hominids from the Upper Pleistocene who, when compared to modern humans, have advanced calcification of the second and third molars, and delay of the canines (Tomkins, 1996). It has been suggested that this variability has a genetic basis, and that the larger jaws of the non-adult hominids allows for earlier eruption of the molars (Bermudez de Castro and Rosas, 2001).

One of the most popular and easy to use macroscopic methods for dental age estimation based on the extent of mineralisation was provided by Schour and Massler (1941). Later Ubelaker (1989) modified this scheme to include larger standard errors that took into account data from his American Indian (Arikara) sample. Some authors have questioned the reliability of this method, as the original data was, in part, taken from an earlier study by Logan and Kronfield (1933) based on 20 children with debilitating illnesses and cleft palate, which may have affected dental development (Smith, 1991). Hence, radiographic standards derived by Moorrees and colleagues (1963a, b) for deciduous and permanent dentition are preferred, as they are based on a larger sample of 246 (deciduous) and 99 (permanent) teeth of White American males and females. Data for the permanent teeth were refined and tabulated by Smith (1991) reducing the inter- and intra-observer error inherent in the complex diagrams produced by Moorrees and colleagues. The method relies on observation of the deciduous canine, first and second deciduous molars and six of the permanent mandibular teeth. Bolaños and colleagues (2000) warn that age estimations based on dental

Table 3.1 *Published modern dental age assessment data for children from different geographical regions*

Population	Source	Sample size	Notes
African	Chagula (1960)	990 males	Eruption of third molars
American (White)	Moorrees *et al.* (1963a, b)	246	Formation stages for deciduous and permanent mandibular teeth
American (Black)	Garn *et al.* (1973)	3868	Permanent tooth emergence in low-income population
American Black, White, Latino	Lovey (1983)	1085	Mineralisation of permanent dentition
American Black American White	Harris and McKee (1990)	335 655	Mineralisation of permanent teeth in males and females
Australian Aborigine Australian, New Zealand White	Fanning and Moorrees (1969)	210	Mineralisation of the permanent molars
Belgian	Willems *et al.* (2001)	2116	Corrected Demirjian *et al.*'s (1973) permanent tooth development technique
Belgian	Gunst *et al.* (2003)	2513	Third molar development
Chinese	Davis and Hägg (1994)	204	Tested Demirjian *et al.*'s (1973) permanent tooth development technique
European	Gustafson and Koch (1974)	ns	Combined published data for deciduous and permanent teeth
French Canadian	Demirjian *et al.* (1973)	2928	Mandibular permanent dentition
French Canadian	Demirjian and Goldstein (1976)	2407 males 2349 females	Updated and simplified method following Demirjian *et al.* (1973)
Indian (Pima)	Dahlberg and Menegaz-Bock (1958)	957	Compared eruption times of permanent dentition
European descent	Nolla (1960)	50 males, 50 females	Mineralisation of permanent teeth in males and females
Mexican American	Nichols *et al.* (1983)	500	Compared with Demirjian *et al.*'s (1973) French Canadian data
Norwegian	Nykänen *et al.* (1998)	261	Tested accuracy of Demirjian and Goldstein (1976) on their sample
Spanish	Bolaños *et al.* (2003)	786	Third molar development; tested Nolla's (1960) data
Zambian	Gillett (1997)	721	Permanent tooth emergence
Various: Bangladeshi Guatemalan Japanese Javanese	Holman and Jones (1998)	397 1271 114 468	Deciduous tooth emergence times

calcification in modern children are *most* precise for those less than 10 years of age, perhaps due to the greater number of stages to be scored and teeth that can be observed. In addition, they suggest the most accurate ages can be gained from the central maxillary incisors, the first mandibular molar and the canine in boys, and the central maxillary incisors and the first and second mandibular molars in girls.

Dental eruption begins as the cleft at the beginning of root formation mineralises and pushes the tooth up through the alveolar bone until it finally occludes with the tooth above or below it (Gleiser and Hunt, 1955). Evidence of clinical eruption, or the emergence of the tooth through the gum, is distinct from the eruption through the alveolar bone seen in the hard tissues, which occurs slightly earlier (Hillson, 1992), and this should be taken into account when choosing ageing standards. Again, the timing and sequence of tooth emergence has been shown to vary between populations and with disease (Dahlberg and Menegaz-Bock, 1958; Niswander, 1965; Garn *et al.*, 1973; Garn and Clark, 1976; Tomkins, 1996; Gillett, 1997), with caries, severe malnutrition or premature shedding of the deciduous teeth reported to delay the eruption of the succeeding permanent dentition (Demirjian, 1990; Larsen, 1997). The third molar, in particular, is highly variable in its time of eruption, and in some cases may become impacted and not erupt at all. Tanguay and colleagues (1986) suggested that standards of emergence for boys and girls would be more precise if based on tooth length-for-age, rather than chronological age, as the length of the tooth at time of emergence varies between the sexes. Liversidge and Molleson (1999) produced such a method for measuring tooth length both macroscopically and from radiographs based on archaeological remains. When this standard was tested for accuracy on a modern sample, ages were underestimated and accuracy decreased with age (Liversidge *et al.*, 2003). Following these studies, Muller-Bolla and colleagues (2003) introduced a simplified method of dental age, using counts of erupted permanent teeth, based on 5848 modern (AD 2000–2001) White children from France. Caution is advised when choosing a method of dental ageing for modern skeletonised remains, when developmental and eruption timings were derived from archaeological samples of unknown age, and vice versa.

For fetal remains, several dental development standards have been published (Schour and Massler, 1941; Moorrees *et al.*, 1963a; Kraus and Jordan, 1965; Lunt and Law, 1974; Deutsch *et al.*, 1985; and more recently Nyström *et al.*, 2000; Nyström and Ranta, 2003). The mineralisation of the deciduous dentition is considered to be more robust to external influences such as socio-economic conditions than the permanent teeth, although ancestral and sex differences still exist (Liversidge *et al.*, 2003). Dental emergence is not considered as reliable; for example, Demirjian (1990) found that all the teeth, with the exception of the first deciduous molar, emerged 1 month earlier in boys than in girls, and Holman and Yamaguchi (2004) have shown that poor and even moderate nutritional stress can delay the emergence of the deciduous teeth.

Estimates of age based on dental wear in non-adult dentition is rarely attempted, and as with adult cases, the degree of attrition is dependent on the type of food eaten and the quality of the dental enamel, and is specific to each sample. The evidence of wear in deciduous teeth indicates occlusion, and implies the introduction of solid foods into the child's diet, which may provide a potential proxy for weaning ages in a sample (Skinner, 1997). Currently, data on dental wear in past populations are only available for adults, with degrees of attrition scored from the youngest to oldest (e.g. Miles, 1963; Scott, 1979; Brothwell, 1981). Modification of these methods on a large non-adult group has yet to be attempted (Saunders, 2000).

Few standards exist for the formation of deciduous crowns based on individuals of known gestational age (Saunders, 2000), and in fetal remains, the developing crowns sit in large open crypts, which often result in the loss of tooth germs during excavation and storage. In some situations, it may be necessary to use macroscopic methods to assign an age using radiographic data, which can produce errors as a higher stage of development is often scored when viewing the tooth itself rather than its image on radiograph. Radiographs will show a slightly later time for the formation of crowns and roots than data based on dissected material, as the dental tissue needs a greater degree of calcification before being visible on a radiograph (Hillson, 1992). For example, the first permanent molar does not show up on a clinical radiograph until 6 months, despite being calcified prenatally (Huda and Bowman, 1995). On a radiograph of dry bone, teeth may rotate in the socket and obscure the true developmental stage of the tooth. Similarly, fragile mineralised material, for instance the cleft at the start of root development, may be damaged post-mortem resulting in a lower score for the macroscopic assessment. One way to combat the errors in interpreting chronological from biological age, and underestimates that may be introduced by damaged teeth and visualisation on radiograph, is to examine microscopic features. Microscopic ageing techniques are based on the incremental markings within the dental microstructure. Unlike macroscopic or radiographic ageing methods, the chronological age does not need to be extrapolated from a set of standards but can be assessed directly, reducing error as a result of inter-population variability (Huda and Bowman, 1995).

Dental growth layers reflect a physiological rhythm and are represented by incremental markers such as cross striations, striae of Retzius and perikymata. Cross striations along the length of the enamel prisms are thought to result from a 24-hour variation in the enamel matrix secretion (Boyde, 1963), whereas the coarser striae of Retzius (seen on the surface as perikymata) represent near-weekly (4–11 days) variation (Huda and Bowman, 1995). The first study on microscopic dental ageing was carried out by Asper (1916) on human permanent canines, and was followed by that of Massler and Schour (1946), who provided a technique for estimating age in days. The method devised by Boyde (1963), originally conducted on two forensic and one archaeological case, was employed by Huda and Bowman (1995) to distinguish more precisely between commingled non-adult remains when macroscopic dental assessment and skeletal development proved inadequate. Smith and Avishai (2005) have advocated the use of the neonatal line as a marker from which subsequent enamel layers can be used to provide a more exact age at death than those provided by diaphyseal length or crown development. They warn that observer experience, decomposition of partially mineralised enamel and variations in the plane of the thin section can affect the accuracy of the technique.

3.1.2 Skeletal estimates of fetal age

When the dentition is unavailable, estimates of age on non-adult remains rely on the development, growth and maturation of the skeleton. Establishing the age of fetal remains is mainly carried out using diaphyseal lengths, and is critical in many forensic cases as a means of assessing the viability of the fetus, and to assist in cases of abortion legislation and infanticide. Errors in the relationship between long-bone diaphyseal lengths and gestational age are well known but difficult to control for, in both forensic and archaeological contexts. The nutritional status of the mother during pregnancy, her weight and her height, all affect the size and weight of the child at birth and hence the length of the diaphyses (Hauspie *et al.*, 1994; Adair, 2004). Additional stress factors such as noise pollution and smoking will decrease size for age (Schell, 1981; Sobrian *et al.*, 1997). It has also been proposed that female fetuses mature earlier than males in both leg length and weight (Lampl and Jeanty, 2003), potentially resulting in longer diaphyseal lengths and older age estimates in female-dominated samples. The issue of shrinkage of the drying out diaphyses in fetal remains was first raised by Rollell in the 1800s (cited in Ingalls, 1927) and may be an additional source of error. Studies of wet to dry bone have shown up to 10% shrinkage, and from wet to burnt bone up to 32% shrinkage in fetuses of 4 lunar months, decreasing to 1% and 2% respectively in newborns (Huxley, 1998; Huxley and Kósa, 1999). However, Warren (1999) reported no significant differences in ages derived from radiographic (wet) diaphyseal lengths compared to dry bone measurements.

Data for age estimates based on fetal diaphyseal lengths have been provided in numerous studies in the past (Balthazard and Dervieux, 1921; Scammon and Calkins, 1923; Olivier and Pineau, 1960; Fazekas and Kósa, 1978; Scheuer *et al.*, 1980; Kósa, 1989). For many years, ages based on data by Fazekas and Kósa (1978) were commonly employed. There have been concerns raised about the validity of the data, which were based on 138 Hungarian perinates of unknown age. A variety of data for ageing fetal remains from diaphyseal lengths have most recently been published by Scheuer and Black (2000). Attempts are also being made to provide updated radiographic data of modern fetal remains using methods that can be applied to dry bone (Warren, 1999; Adaline *et al.*, 2001; Piercecchi-Marti *et al.*, 2002), but caution is advised when attempting to age fetuses with known or suspected congenital pathology (Sherwood *et al.*, 2000). Huxley and Angevine (1998) provide a conversion chart for the transfer of ages reported in lunar months (anthropological sources) or gestational weeks (clinical sources) to allow viability of the fetus to be more accurately assessed.

General assessments of age for perinates (whether pre- or postnatal) have been devised based on the appearance, size and fusion of growth centres. These include the temporal ring and plate (Scheuer and Black, 2000), the pars petrosa (Way, 2003) and the suture mendosa and pars basilaris of the occipital (Scheuer and Maclaughlin-Black, 1994). For example, the pars basilaris was originally measured and described in detail by Redfield (1970) and later by Fazekas and Kósa (1978). Measurements of the maximum length (ML), sagittal length (SL) and maximum width (W) of the pars basilaris enabled a determination of the age of the fetus or infant. In 1994, Scheuer and Maclaughlin-Black tested this method on both archaeological and known-age individuals from nineteenth-century England. They concluded that if the pars basilaris was longer than it was wide the individual was less than 28 weeks in utero, and if the width was greater than the length the individual was over 5 months of age. They suggested that the maximum width and sagittal length were of the greatest value when distinguishing between an early or late fetus, and the maximum width and maximum length were more accurate when identifying early or late infant material (Scheuer and Maclaughlin-Black, 1994). Tocheri and Molto (2002) showed age assessments based on the pars basilaris and those attained for femoral diaphyseal length and dental development agreed in 87% of cases in a sample from Dakhleh. This bone is particularly useful in the age assessment of fetal remains because its compact and robust structure means it is often recovered intact. More recently, new data based on the dimensions of the fetal atlas and axis (Castellana and Kósa, 2001), vertebral column (Kósa and Castellana, 2005) and the mandibular ramus (Norris, 2002) have been provided in the forensic literature.

3.1.3   Skeletal development and maturation

When the teeth are absent or apical closure is complete, age assessments of older children and adolescents become based on diaphyseal lengths, and the appearance and fusion of the secondary growth centres (epiphyses) to the diaphyses. Ubelaker (1989) studied the long-bone lengths of the Arikara Proto-historic sample from Dakota, USA, which he initially aged using the dental calcification chart devised by Moorrees and co-workers (1963a,b). The measurements were then divided into 1-year age categories up to the age of 18.5 years. As the skeletal development of a Native American population is delayed when compared to that of a White population, Ubelaker added a large standard deviation (±2 years) and a range of variation. However, in some cases there are no individuals to represent a particular age category (i.e. 8.5–9.5, 13.5–14.5 and 16.5–17.5 years) and other age categories are constructed using only one individual. Hence, this method is considered most reliable for assigning an age to those between birth and 2.5 years, where a greater amount of data exists. Data for the diaphyseal lengths of older children are also available from modern studies (Maresh, 1955; Anderson *et al.*, 1964; Hoffman, 1979; Sundick, 1978; Hunt and Hatch, 1981; Scheuer and Black, 2000). Hoffman (1979) and Ubelaker (1974) argue that of all the long bones, the femoral diaphyseal length is the most reliable indicator of age. Nevertheless, diaphyseal lengths are known to be affected by external factors such as malnutrition and infection. For this reason, these measures are normally employed to identify physiological stress in an individual through growth profiles, which compare bone length to dental age (see Chapter 4), rather than to provide chronological age. As the child ages and the epiphyses begin to fuse to the diaphyses, long-bone measurements are abandoned in favour of maturational assessments.

In the postnatal growth period, epiphyses begin to fuse when elongation of the diaphyseal shaft occurs at the expense of the epiphyseal cartilage, until it is completely replaced with bone. Fusion of the epiphyses begins between the ages of 11 and 12 years at the elbow (distal humerus and proximal radius) and ends with fusion at the knee (distal femur, proximal tibia and fibula) which is generally complete at 17 years in females and 19 years in males. The nature of epiphyseal closure was first outlined by Hasselwander (1902) and normally begins with fusion at the central aspect of the growth plate, with union proceeding centrifugally until it is complete (Dvonch and Bunch, 1983). There are several exceptions to this rule, with the proximal femur showing initial fusion at the superior margin and continuing inferiorly, and the proximal tibia first exhibiting union from the anterior–medial aspect to the posterior margin (Dvonch and Bunch, 1983). In addition, Albert and colleagues (Albert and Greene, 1999; Albert *et al.*, 2001) have suggested that environmental stress results in asymmetrical epiphyseal fusion, and hence asymmetric long-bone length, potentially providing two different age assessments in stressed individuals.

As the last region of the growth plate to become completely ossified is usually at the bone margins, in dry bone a groove may still be observed around the circumference (Roche, 1978). This is known as partial fusion. When the groove has completely disappeared, complete fusion is recorded. The ossified margin between the old epiphysis and diaphysis may still be visible on a radiograph for a number of years after the macroscopic groove has disappeared. For example, Cope (1946) reported the appearance of the fusion line on the distal femur and proximal tibia in an individual over 70 years of age! Therefore, standards based on macroscopic assessment may record complete fusion at an earlier age than a radiographic study. Post-mortem detachment of a newly fused epiphysis often occurs, and can be distinguished from non-union by subtle damage to the metaphysis (usually centrally), which exposes the underlying trabecular bone, and the lack of an undulated and billowing surface characteristic of an unfused epiphyseal surface.

Data for establishing chronological age based on skeletal maturation appear in many general osteological texts, and stem from skeletal and radiological studies carried out in the early 1900s, set to redress the discrepancies evident in anatomical texts of the day (Stevenson, 1924; Davies and Parsons, 1927; Paterson, 1929; Todd, 1930; Flecker, 1932, 1942; Acheson, 1957). A variety of ages can be assigned, depending on the bone or standard being used. Most standards require the observer to score whether the growth plate is open, partially fused or completely fused (e.g. Buikstra and Ubelaker, 1994), with the sequence of fusion beginning at the elbow through to the hip, ankle, knee, wrist and finally the shoulder. Once fusion of the long bones is complete, late-fusing epiphyses, such as the basilar–occipital synchondrosis, are used for ageing older individuals, but recent research on modern cadavers has indicated it is only reliable for females (Kahana *et al.*, 2003), making this a difficult feature to use in individuals of unknown sex. Other methods, such as the fusion of the thoracic and lumbar vertebral rings in males and females, have helped to refine our ageing methods in adolescents (Albert and Maples, 1995; Albert, 1998; Albert and McCallister, 2004). In forensic contexts, adolescent deaths may occur due to teenage suicides or warfare, or the young people may be vulnerable if they are living on the streets. In the latter case, they are likely to be malnourished or neglected, resulting in a potential delay in their skeletal maturation. Nevertheless, useful biological markers in adolescence include the fusion of the hook of the hamate, capping of the epiphysis of the third metacarpal and, in White children, the completion of the canine root (but not apical closure). All of these features are thought to signify the beginning of the growth spurt and puberty (Chertkow, 1980).

In general, the maturation of girls is recognised to be 2 years in advance of that of boys, and this advancement is increased in obese children (Mellits *et al.*, 1971). As with dental ageing, differences between populations are also evident. For example, Pakistani and American Black children tend to mature earlier than White Americans (Garn and Clark, 1976; Rikhasor *et al.*, 1999). Although socio-economic status has the greatest impact on maturation times, whatever the ancestry (Schmeling *et al.*, 2000), there are few studies available that examine epiphyseal fusion rates in large samples of children of different ancestry, from good economic backgrounds, that would allow the assessment of the true differences to be carried out. In forensic pathology, the most common clinical ageing methods are based on the appearance of secondary ossification centres of the hand and foot using the Greulich–Pyle *Radiographic Atlas* (Greulich and Pyle, 1959) or Tanner Whitehouse System (Tanner *et al.*, 1975). Unfortunately, these techniques are not reliable for skeletonised remains where tiny epiphyses are rarely recovered, although the method may be of use in partially decomposed remains or cadavers. Other skeletal indicators, such as the appearance of ossification centres and fusion of the various segments of the spine, may aid age estimation in cadavers of forensic cases when all the elements are known to be present. Precise detail on the timing of appearance and fusion of various elements in the developing skeleton can be obtained from either of the two excellent texts devoted entirely to non-adult osteology by Scheuer and Black (2000, 2004).

3.2   Sex determination

Sex is normally determined in the adult skeleton using morphological features of the skull and pelvis, and measurement of the long bones. Sexual dimorphism exists due to the release of hormones during the development of the fetus. If cells are not masculinised by the release of testosterone in the 10th week of gestation, a fetus will develop along ovarian (female) lines (Saunders, 2000). Unfortunately, the sexually dimorphic characteristics of the pelvis and skull do not become apparent until puberty, meaning that assigning sex to non-adult skeletons is more problematic. In addition, chemical differences between male and female skeletal remains, such as different levels of citrate or calcium, phosphorous and strontium, are only recognised in males and females of reproductive age (Dennison, 1979; Beattie, 1982). However, some morphological characteristics, such as the greater sciatic notch, are often so different in males and females that researchers believe that they must already be present in children.

In biological and forensic anthropology we determine *biological sex*. This term is not to be confused with *gender*, a cultural construct that refers to the social significance placed upon males and females in society (Mays and Cox, 2000). As with biological age and the concept of childhood, biological sex is the reference point from which inferences about gender in past society can be made. Sex estimation is the 'holy grail' of non-adult osteology. As females grow and mature faster than males, accurate sex estimation of non-adults would allow for age estimates to be refined and the social treatment of males and females at birth and during childhood to be examined. Sex identification of fetuses would be used to answer questions about changing sex ratios and their predisposing factors (e.g. parental investment and economic status (Koziel and Ulijaszek, 2001) and the fragility of male fetuses). The proportion of male to female children displaying various degrees of trauma and pathology could tell us much about their preferential treatment and life-course, their susceptibility to disease, handling in society through burial rites and grave inclusions, division of labour, and at what age boys and girls became accepted as adult members of a society. For example, in 1997 Rega assessed the relationship between gravegoods, sex and gender in skeletons from a Bronze Age cemetery at Mokrin in former Yugoslavia and assigned the sex of children using canine tooth crown measurements, their body orientation and grave goods. Rega (1997) illustrated that young females were afforded the same burial rites and gender identity as the adult females, but the large proportion of infant remains sexed as female fell outside the normal sex ratio of 1 : 1, suggesting that males were assigned female gender, and that only after 17 years of age did male burials become associated with male gendered knives.

Unfortunately, sex estimations from non-adult skeletons are notoriously difficult, making associated anthropological techniques unreliable, or forcing us to add greater error ranges to our results in order to account for unknown sex. Most anthropological techniques used to assign sex to non-adults focus on the size and morphological traits of the dentition, cranium, mandible and pelvis. When all cranial and postcranial traits are used, sex determination in adult skeletons can provide an accuracy of 98–100% (Buikstra and Mielke, 1985), and in general, 95% accuracy is expected from anthropological techniques used in forensic cases. In children, most methods fail to yield an accuracy of 70%, with the degree of overlap between male and female traits and problems with reproducing accurate results on different populations continuing to make this a challenging area of investigation (Loth and Henneberg, 2001; Scheuer, 2002).

3.2.1   *Sexing from the dentition*

Measurements of the deciduous dentition have shown that teeth of boys are generally larger than those of girls (Black, 1978; De Vito and Saunders, 1990) yielding an accuracy between 64% and 90%. Black (1978) developed discriminant function analyses for deciduous dentitions resulting in an accuracy of 75%, although he found the sexual dimorphism to be less marked in the deciduous teeth than in the permanent dentition. In primates, the canines show the greatest degree of sexual dimorphism, with up to a 50% difference in the size of the mesiodistal crown diameter in a male compared to a female (Garn *et al.*, 1967). In humans, the canine shows its greatest difference between boys and girls between the ages of 5 and 12 years (Bailit and Hunt, 1964), with male canines over 4% larger than female canines (Keene, 1998). This degree of sexual dimorphism varies between populations (Garn *et al.*, 1967). Thompson and colleagues (1975) reported that the further away the teeth were from the canine, the less variability there was between males and females. Moss and Moss-Salentijn (1977) suggest that this canine sexual dimorphism in crown size is due to the thickness of the enamel, as amelogenesis continues for a longer time in males than in females. However, the large degree of overlap between males and females caused research into canine dimensions to wane. In 1990, fresh attempts were made to assess the accuracy of sexing children based on the dimensions of the deciduous teeth. De Vito and Saunders (1990) achieved the highest reported accuracy (76–90%), although no tooth over any other showed the most marked differences. The authors acknowledged that the degree of sexual dimorphism in their Burlington (Vermont, USA) group was particularly marked compared to studies of other North American, Australian or Icelandic groups.

It is important to bear in mind the unknown genetic and environmental factors that may affect the accuracy of sex estimates based on crown dimensions. For example, mothers with diabetes and hyperthyroidism can produce offspring with larger tooth dimensions, while maternal hypertension and low birth weight will result in smaller dental crowns for both the deciduous and permanent teeth (Garn *et al.*, 1979). Further research into prematurity and crown dimensions showed a less predictable pattern in the permanent teeth, with preterm White males and Black females showing increased dimensions (Harila-Kaera *et al.*, 2001), while in the deciduous teeth, males have consistently larger teeth with prematurity than females (Harila *et al.*, 2003).

Hunt and Gleiser (1955) proposed a method of sexing non-adults using correlations between dental and skeletal age, based on the principle that males mature skeletally more slowly than females, while the rate of dental development between the sexes is similar. That is to say that, in an unknown child, if the ages attained independently for dental and skeletal age are closely correlated using male standards, but divergent when compared to female standards, then the child is male. This technique was based on the development of wrist bones from clinical radiographs, but these tiny epiphyses are susceptible to taphonomic processes and are often not recovered in skeletonised remains, although the method may be of use in partially skeletonised individuals or cadavers. Again, the children that enter the forensic record are often malnourished or

neglected, resulting in a delay in their maturation, whatever their sex, in relation to their dental development (Eveleth and Tanner, 1990). Following on from
Hunt and Gleiser (1955), Bailit and Hunt (1964) tested whether the mandibular canine in particular was an effective estimator of sex when chronological
age was already known, as a canine belonging to a male should more closely
approximate known age based on male standards than the canine of a female.
This method resulted in accurate sex estimation in 70% of cases, compared to
only 58% accuracy when sexing the child based on Hunt and Gleiser's (1955)
method. Hence, sex estimations based on the canine are more reliable when the
chronological age of the child is already known, a rare event in archaeological
or forensic specimens.

3.2.2   Sexing from skeletal morphology: skull

In more recent years, anthropologists have shifted their focus away from the dentition to the sexual dimorphism evident in the non-adult cranial and postcranial
skeleton. Such studies have been encouraged by the acquisition of archaeological material including children of known age and sex. In 1993, Schutkowski
examined the extent of sexual dimorphism in the mandible and ilium of 37
boys and 24 girls aged between 0 and 5 years from Christ Church Spitalfields
in London, and reported an accuracy of between 70% and 90% (but see Halcomb
and Konigsberg, 1995). For the first time mandibular morphology was used as
a sexing criteria for non-adults. However, when Schutkowski's mandibular features were tested by Loth (1996) accuracy rates for chin morphology were only
33%, gonial eversion was 37% and the dental arcade shape was not considered
sexually dimorphic at all (Loth and Henneberg, 2001). In the same year, Loth
and Henneberg (1996) demonstrated adult differences in chin shape (rounded
in females, squared off in males) were evident in children from the age of 6
years, and were identical to adult forms by 13 years, with the best predictive
accuracy of 69% occurring between the ages of 6 and 19 years.

In 1998, Molleson and colleagues again attempted to assign sex to non-
adults from Christ Church Spitalfields, as well as an undocumented sample
from Wharram Percy in Yorkshire, using cranial characteristics which included
the mandibular features (ramus angle and chin). They expanded their criteria
to include the orbit, already recognised in the adult facial skeleton as showing
characteristic male and female traits. Their method involved scoring each feature from -2 to $+2$, with -2 being hyperfeminine, 0 indeterminate and $+2$
hypermasculine. These scores were then summed in each specimen to give a
combined facial score (CFS). At Spitalfields they were able to determine the
correct sex of the child in 89% of cases (Molleson *et al.*, 1998). The need to have

all three features preserved limited the number of individuals on which these features could be tested to only 18. The orbital features were most commonly found to be missing, and the CFS showed a preference for sexing females as male (Fig. 3.1). Despite being one of the only documented non-adult collections in the UK, Spitalfields has a low proportion of females and has a small known-age sample, making it less than ideal for this type of analysis. The mandible once again received attention as a potential sex indicator in 2001, when Loth and Henneberg reported differences in the morphology of the mandible in non-adults from the documented South African Dart Collection in Pretoria. They reported 81% accuracy in sex estimation based on the morphology of the inferior border and body of the mandible, with females displaying a rounded gradual border, compared to the male straight lateral border and angulation at the point of the canines (Loth and Henneberg, 2001). When tested on a limited number of French and South African children, the accuracy was reported at 82%. Scheuer (2002) blind-tested this method on the Spitalfields collection but reported an accuracy of only 64%.

Coussens and colleagues (2002) used Loth and Henneberg's (2001) mandibular criteria to examine the differences in robusticity index of the femur and humerus in children aged between 0 and 4 years from St Mary's Anglican Church in South Australia (AD 1847–1925). Their preliminary results suggest a correlation between diaphyseal robusticity and sex that warrants further investigation in known-sex samples. Sutter (2003) tested all cranial and postcranial criteria on prehistoric known-sex non-adult mummies from northern Chile and, although all traits with the exception of the gonial angle showed some level of sexual dimorphism, only the sciatic notch depth and dental arch criteria produced acceptable levels of accuracy at 81.5%, in individuals between the ages of 0 and 5 years. A common conclusion in studies that have tested the accuracy of morphological methods for sexing non-adults is that many features are biased towards sexing males, and that at certain ages these features become less reliable, suggesting that growth has a bigger impact on morphology than sex (Hunt, 1990; Schutkowski, 1993; Loth and Henneberg, 2001; Nicholls, 2003).

3.2.3 *Sexing from skeletal morphology: pelvis*

Studies into the morphological differences between male and female fetal pelvises began as early as 1876, with differences in the sub-pubic angle, pelvic inlet and sciatic notch noted in articulated anatomical specimens from as early as the 3rd month of gestation (von Fehling, 1876; Thomson, 1899). Reynolds (1945, 1947) suggested that males demonstrated greater ilium breadth compared to the greater sciatic notch breadth and pubis length in females, in both the fetal

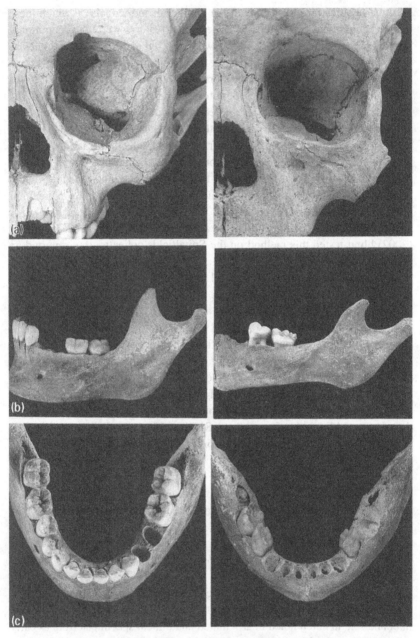

Figure 3.1 Criteria for sexing non-adult skulls. With +2 scores (hypermasculine) on
the left and −2 scores (hyperfeminine) on the right side showing the contrasting
morphology of the orbit (a), mandibular angle (b) and the mentum (c) in non-adults.
From Molleson and colleagues (1998:721), reproduced with kind permission from the
authors and Elsevier.

and 2–9-year age categories, with females showing a greater variability in measurements. Reynolds (1947) also noted that the differences between the sexes became less distinct in the older age categories. Nevertheless, these methods were successfully used to sex a child in one of the earliest forensic identification cases involving a non-adult in the UK (Imrie and Wyburn, 1958). Boucher (1955, 1957) demonstrated differences in the male and female sub-pubic angle in both Black and White fetal samples, showing similar degrees of difference between the fetuses as seen in adults from the same populations. These angles were produced by the tripartite cartilage that holds the pelvis together in life, but which is not present post-mortem. The reliability of the ischium length and diameter of the acetabular surface in non-adults was examined most recently by Rissech and colleagues (2003) on a sample of 327 innominates from St Bride's in London and the Lisbon Collection. They reported significant differences in the acetabular index (vertical diameter and horizontal diameter of the unfused ischium acetabular surface) in males and females aged between 5 and 9 years. They also suggested that adult-type sexual differences in pubis length are discernible by 15–16 years, and that differences in acetabular surface measurements are apparent by 12 years.

The sciatic notch is one of the most visible sex indicators in the adult pelvis, and has received much of the attention in non-adult sex studies. Results are often contradictory, with some studies showing females to have higher sciatic notch index (depth over breadth) than males (Boucher, 1957) and others suggesting this indicator is less dependable (Weaver, 1980). Holcomb and Konigsberg (1995) reasoned that much of the difficulty with the reliability of this indicator lies in establishing the appropriate landmarks, and so they developed a digital method to plot mathematically the differences between male and female fetal specimens. Nevertheless, the authors did not report any significant differences between males and females in the 133 fetal ilia from the Trotter Collection, and sex was predicted correctly in only 60% of cases compared to Schutkowski's (1993) previously reported 95% accuracy.

In 1980, Weaver introduced auricular surface elevation as a potential trait in fetal and infant ilia, where: 'if the sacro-iliac surface was elevated from the ilium along its entire length and along both the anterior and posterior edges of the sacro-iliac surface, the auricular surface was considered elevated.' (Weaver, 1980:192). Here, an elevated auricular surface indicates a female whereas a non-elevated surface indicates a male. Weaver reported this trait to predict sex accurately in 75% of females, and 92% of males. Hunt (1990) argued that this feature was more likely to be related to age and growth of the pelvis as it produced unrealistic sex ratios when applied to an archaeological sample. This theory was, in part, supported by Mittler and Sheridan (1992) who examined the trait in 58 children between the ages of 0 and 18 years. Although the method correctly sexed males in 85% of cases, it only provided a correct sex estimate for females in 58% of cases, suggesting that 'entire' elevation is difficult to score. Mittler and Sheridan (1992) argue that in early childhood, most auricular surfaces are flat, only becoming raised in females after 9 years of age, and even then, this feature failed to develop in 26% of their female cases. Mittler and Sheridan (1992) conclude that in archaeological studies, auricular surface elevation may be of limited use and has a male bias. In forensic cases concerning children older than 9 years, an elevated auricular surface would probably indicate a female. A later study on pelvic traits by Sutter (2003) found similar accuracy rates for males (77.8%) and females (60.9%) based on known-sex mummies from Chile, but reported that the arch criteria showed the best results in children with pooled ages (82%). For those under 5 years, the arch and depth of the sciatic notch performed equally well (81.5%). It is debatable whether even these levels of accuracy are good enough for a court of law.

3.2.4   Sexing using DNA analysis

In the late 1990s there was an increase in the use of DNA sex typing for archaeological child remains or fragmentary adult skeletal remains. Many of these studies focussed on the sex of infants thought to be the victims of infanticide (Faerman *et al.*, 1997, 1998; Waldron *et al.*, 1999; Mays and Faerman, 2001), although some studies have used DNA analyses to sex older children (Capellini *et al.*, 2004). The differential amplification of the X-specific and Y-specific sequences in ancient bone can mean males will be wrongly identified as female if the X-specific sequence is preferentially amplified over the shorter, more easily degraded Y chromosome, and caution in the type of technique used is essential, e.g. employing the Dot Blot test to confirm results (Colson *et al.*, 1997; Palmirotta *et al.*, 1997; Saunders and Yang, 1999; Michael and Brauner, 2004). In addition, the choice of bone sample can influence the success of the analysis, with Faerman and colleagues (1995) reporting that DNA was better preserved in the teeth, cranial bones and cortices of long bones than in the ribs.

In 2000, Cunha and colleagues attempted to sex the remains of two 11-year-old children recovered from a medieval convent in Portugal. Although not successful for one of the children, ancient DNA (aDNA) results for the second child suggested it was a boy, raising questions about the true exclusivity of the convent, where male burials were documented to be prohibited. Although the authors suggest they sexed the non-adult remains first using skeletal traits, no comparison of the result was made (Cunha *et al.*, 2000). The development of studies that link morphological traits with aDNA analysis (e.g. Waldron *et al.*,

1999) must continue if we are to refine our techniques in those cases where DNA analysis is not reliable. As yet, aDNA has not been amplified for known male and female non-adult skeletal series with the expense of testing samples hindering any large-scale studies, while we still struggle with problems of contamination, consistency, control, false negatives and false positives. In forensic cases, where soft tissue or body fluids remain, forensic pathologists employ rapid laboratory testing using clinical chemistry analysis for testosterone, oestradiol and cortisol as an early, but not sole, indicator of sex. In addition, total human chorionic gonadotrophin may be used to assess whether an individual was pregnant. Caution is advised in the interpretation of the results when the individual is pubertal, as hormonal imbalances may result in false positive identification of females who are in fact males (Lewis and Rutty, 2003).

3.3   Ancestry

Assessments of ancestry are rarely carried out in archaeological research, and this determination is usually confined to forensic anthropology, where it provides one of the main biological profiles for an unknown individual (along with age, sex and stature). One of the major problems with ancestry assessment, particularly based on the morphology of the skull, is that people rarely fall within the three traditional groups (i.e. Caucasian, Negroid and Mongoloid) and many show a range of morphological characteristics reflective of the high degree of genetic admixing in modern populations. The biological affinity assigned to an individual may be different to that perceived by the people who knew the individual, or the individual themselves, with some people of mixed parentage being seen as either Black or White, depending on who is describing them. In addition, skeletal morphological characteristics are not useful in distinguishing Italians from Norwegians, for instance, where differences in appearance are usually based on hair and eye colour, and are not expressed skeletally. Nevertheless, ancestry assessments are an important part of the forensic anthropologist's work, and assessments are additionally carried out using dental traits and, most recently, chemical analysis.

Estimates of geographical affiliation based on skeletal traits in infants and children are problematic. Many of the features identified in the skull are secondary sexual characteristics that occur with the continued development of the jaw and sinuses at puberty (Briggs, 1998); but few samples of known-ancestry non-adult material exist in order for us to develop these criteria (St Hoyme and Iscan, 1989). Recent research by Weinberg and colleagues (2002) demonstrated the potential for cranial morphology to provide an ancestry assessment in African and European derived fetuses. Their initial studies indicate that

European fetuses have narrow supraoccipital portions, more circular temporal squama, longer and lower vomers, more prominent nasal spines and more pronounced nasal sills than their African counterparts. Dental arch dimensions in Black children have been shown to be significantly larger than in White children (Harris, 2001). Warren and colleagues (2002) have suggested a difference in long-bone allometry between African-American and European-American fetuses, suggesting a genetic rather than environmental basis for the phenomenon. In the archaeological context, Özbek (2001) illustrated some remarkable cases of cranial modification in 13 children from Değirmentepe in Chalcolithic Turkey. Children aged between 1 month and 14 years demonstrated compression over the frontal bones, parietals and occipital as their heads were elongated anteroposteriorly by the use of first one, and then two bandages. Özbek argues that such practices provide evidence for the impact of long-distance trade development, and the need for a group to maintain overt ethnic markers. Pathological conditions, such as those related to sickle-cell anaemia and thalassaemia, may indicate a child of Mediterranean or African descent (Ortner, 2003). For example, Shaw and Bohrer (1979) noted the incidence of ivory epiphyses (stunted radio-opaque epiphyseal plates with absent trabecular structure) was higher in the hand radiographs of Nigerian than in African-American children as the result of sickle-cell anaemia. Cleft-lip-and-palate is also more common in children of Asian ancestry than of European descent (Bowers, 2005).

Research into the computer-simulated geometry of the mandible and the facial bones is being developed and should aid the assessment of ancestry in non-adults (Strand Vidarsdottir *et al.*, 2002; Strand Vidarsdottir, 2003). For example, Buck and Strand Vidarsdottir (2004) carried out an analysis of 174 complete mandibles of adults and children from five distinct geographical groups; African-Americans, Native Americans, White Europeans, Inuit and Pacific Islanders. Results showed a 73% accuracy using the mandibular ramus, and 67.2% accuracy using the mandibular corpus. While this is promising, the authors were unable to take sexual dimorphism in mandibular morphology into account, a factor that has been shown to vary between children of different age and sex in varying degrees throughout the growth period.

Schultz (1923) identified differences in the height and breadth of the nose in 623 Black and White fetuses from the Carnegie Laboratory of Embryology in Washington. Noses were shorter and broader in Black fetuses reflecting similar differences in adults. Tobias (1958) suggested that population affinity differences observed in adult crania are visible by 5 or 6 years of age. To test this theory, Steyn and Henneberg (1997) examined the cranial dimensions of 16 preserved skulls of children from K2 and Mapungubwe thought to be of Black South African origin. They noted reduced cranial breadth characteristics of the adults in the group to be evident in the children from the age of 5 years, compared to the broad heads of Polish children, and the intermediate breadth of American children. However, they caution that the sample size was small, and that the sexes were pooled.

There is also potential for metric and morphological traits to provide an indicator of ancestry from the deciduous and permanent dentition in both archaeological and forensic samples. These should be evident in the dentition of a child from about 9 months of age, and will not remodel. Harris and colleagues (2001) have shown that deciduous molar dimensions are greater in American Blacks than in Whites. This difference was also noted in the proportion of enamel and dentine thickness and the size of the pulp cavity, with Black Americans displaying disproportionately thicker enamel than Whites. Shovel-shaped incisors occur more frequently in the permanent teeth of Asians and Native Americans (90%) and less frequently in Whites and Blacks (less than 15%) (Mizoguichi, 1985), hypodontia is perhaps more common in children from the Orient (Rosenzweig and Garbarski, 1965) and Carabelli's cusps are more common in Europeans (75–85%), Asians and Africans, respectively (Hillson, 1996). These dental epigenetic traits have also been classified for the deciduous dentition (see illustrations in Hanihara, 1961). Lease and Sciulli (2005) used deciduous dental metrics and dental non-metric traits to distinguish between modern European-American and African-American children, based on the incisors, canines and premolars. They concluded that a combination of dental metrics and morphological characteristics of the deciduous teeth could correctly assign children to their ancestral groups in 90–93% of cases. Danforth and Jacobi (2003) warned that not all dental traits appear in both the deciduous and permanent dentition in the same individual, and this was particularly true for the Carabelli's cusp. In addition, Edgar and Lease (2005) have shown that while deciduous and permanent dental traits may be correlated, they may vary in strength of expression.

Snow and Luke (1970) looked for negative evidence to rule out an Indian ancestry in the Oklahoma forensic case, where a slightly developed Carabelli's cusp indicated a White individual, a fact they supported by the lack of shovel-shaped incisors. Snow and Luke were attempting to *confirm* an identity and this type of reasoning is not recommended for unknown cases. If hair survives, then the hair style and transverse morphology of the hair shafts, including melanin distribution, can provide evidence for ancestry (Deedrick, 2000). In modern cases, skin colour from a decomposing body cannot be relied upon as dark skin can become bleached if left lying in the damp for any period of time, and skin exposed to the air turns dark during putrefaction (Byers, 2002:168). Dark-stained bodies in water can lose their entire epidermis, including the pigmented basal layer, thus appearing White (Rutty, 2001). In decomposing, burnt or almost skeletonised remains, melanin pigmentation of the arachnoid of the medulla oblongata can be an indicator of a non-White ancestry (Spitz, 1993).

In cases of mixed ancestry, anthropological techniques are of limited use, and one of the most recent developments in biological and forensic anthropology is the use of stable isotope analysis as a tool in determining the birthplace (from tooth enamel) and more recent geographic mobility (from cortical bone) of an individual (Beard and Johnson, 2000). This technique, known as 'isotope fingerprinting', uses elements such as strontium, aluminium and cadmium to narrow down a geographical origin of the remains and was most famously used in the 'Adam' case where the torso of an African child was recovered from the Thames in London. It has also been proposed that DNA may be used to assess allele frequencies to define social and ethnic group differences. These genetic profiles can then be used to assist in the identification of an individual's ancestry (Richards, 2001). Montgomery and colleagues (2000) used lead and strontium isotopes to plot the movements of one woman and three children aged 5, 8 and 9 years, buried within a Neolithic (mid to late fourth millennium BC) henge monument at Monkton-upon-Wimborne in Dorset, England. Strontium and lead signatures from the dental enamel of the deciduous and permanent dentition suggested that they moved, perhaps as a group, a distance of about 80 km from chalk land to a more radiogenic area, and back again, reflecting for the first time, medium-distance movements of the Neolithic people in this area.

3.4   Summary

Of the three biological identifiers discussed above, age based on dental development is the most reliable of the assessments. Age estimates form the basis of mortality and growth studies in archaeological populations, but are hindered by the continual problem of establishing a reliable method for sexing non-adult remains. Sex determinations based on the mandible are beginning to yield accuracy results as high as 82%, and those based on deciduous teeth dimensions can yield between 76% and 90%. But studies are consistently showing a male bias in the features analysed, and cut-off points in the age at which these features are no longer sexually dimorphic. Just as with sex estimates, the identification of the ancestry of a child is also influenced by morphological changes during growth and at puberty. In both biological and forensic anthropology, the development of isotopic fingerprinting to identify the country of origin of unknown individuals has become a significant tool that may in time replace the traditional and problematic methods of ancestry assessment, particularly when tracing the geographical origins of non-adults who have yet to develop morphological ancestry traits.

Despite our confidence in our ageing criteria, Lampl and Johnston (1996) warned that age-at-death estimates for deceased archaeological individuals based on standards from modern healthy children would result in considerable errors. They aged modern children using the standards commonly employed in bioarhaeology, and found that the children were under-aged by as much as 4 years for skeletal maturation and 3.5 years for dental development. Greater errors may be expected when assigning ages to ancient skeletal samples, due to the confounding environmental stresses suffered by the archaeological populations. In addition, variability between archaeological samples from different periods and environments may result in less than accurate comparisons for age-at-death and, hence, mortality profiles. Using large age-at-death categories for comparative studies may help to reduce this margin of error.

4 Growth and development

4.1 Introduction

Growth is a highly regulated process controlled by the endocrine system. Postnatal growth begins rapidly and gradually slows and stabilises at around 3 years of age. At puberty there is another episode of growth acceleration which, after a period of peak velocity, slows until the epiphyseal ends of the long bones fuse and growth ceases (Karlberg, 1998). The final growth outcome of an individual is the result of a complex interaction between genetic and environmental factors. Secular trends, showing a systematic increase in stature between generations, indicate that improvements in nutrition and healthcare have enabled populations to reach ever greater proportions of their genetic potential (Henneberg, 1997). The physical growth and development of children is a sensitive indicator of the quality of the social, economic and political environment in which they live. For this reason child growth standards are regularly used as measures of the general health status of the overall community, where poor growth is taken as an indicator of unfavourable conditions (Johnston and Zimmer, 1989).

Growth studies are among the most popular and widely published areas of investigation carried out on child remains in bioarchaeology, with an abundance of excellent texts on the subject (Tanner, 1978; Bogin, 1988a; Eveleth and Tanner, 1990; Ulijaszek *et al.*, 1998; Hoppa and Fitzgerald, 1999). Studies of past childhood growth have been used to provide valuable information on nutritional stress, secular trends, prolonged skeletal growth and delayed maturation between contrasting archaeological groups (e.g. hunter–gatherers and agriculturalists, urban and rural), and when compared to modern growth data. The majority of these studies involve the estimation of longitudinal growth of the long bones in relation to dental development. More recent studies have begun to include the cranium and transverse growth (i.e. bone density) of child remains to examine issues of stress and adaptation in past populations.

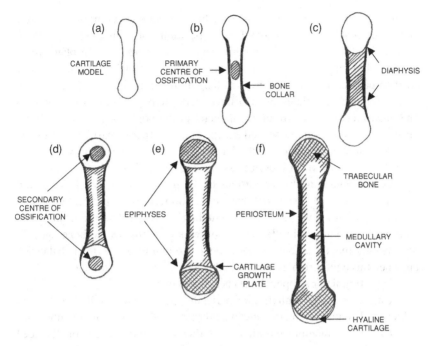

Figure 4.1 Sequence of embryonic bone development.

4.2   Skeletal development and ossification

The development of the skeleton begins around the 12th week of gestation, and involves both intramembranous (within membrane) and endochondral (within cartilage) ossification. Non-adult long bones are made up of a shaft (diaphysis), with a growing zone at the proximal and distal ends (metaphyses), and additional bone segments (epiphyses) that are separated from the metaphyses by a cartilaginous growth plate. The skeleton originates from connective tissue and shares its mesodermal origin with muscular and vascular tissue, illustrating the skeleton's close connection with these systems. The earliest stage of endochondral osteogenesis involves the development of a cartilaginous model that will act as a template for the long-bone shafts. This template is surrounded by the perichondrium (precursor to the periosteum) and, as the model grows, central cells swell and blood vessels form to produce the primary centre of ossification near the centre of the diaphysis, contained within a collar of bone (Fig. 4.1). Osteoblasts are released and begin to deposit osteoid, the precursor to mineralised bone, to form the bone shaft (Steiniche and Hauge, 2003). This process continues along the shaft as blood vessels form at the ends of the model to produce the secondary centres of ossification, or epiphyses.

Throughout the growth process, long bones achieve adult proportions by a combination of longitudinal growth and appositional or circumferential growth. Longitudinal growth is endochondral, and occurs at the growth plate where chondroblasts continually lay down cartilage cells, which gradually become ossified causing the bone to grow in length. Towards the end of the growth period, this cartilage plate begins to thin until the epiphyses fuse to the diaphysis and growth ceases. Growth in the diameter of the bone, or 'bone modelling', occurs as new bone is deposited on the external (periosteal) surface of the shaft by osteoblasts which line the inner layer of the periosteum. At the same time, bone is removed by osteoclasts on the internal (endosteal) surface of the shaft, maintaining the proportions of the medullary cavity (Schönau and Rauch, 2003). Bone tissue continues to be removed and replaced throughout life in a process known as 'bone remodelling' (Frost, 1964), producing more numerous and intercutting Haversian systems (osteons) as the bone ages. In healthy children, the amount of bone added and removed is perfectly balanced, but when this equilibrium is disrupted, the features indicative of bone pathology (e.g. hypertrophic or atrophic bone) become evident.

The flat bones of the skull (frontal, parietals), clavicle and maxilla form within a thickened connective membrane in a process referred to as intramembranous ossification. The membrane is directly ossified by osteoblasts without the need for a cartilage precursor, allowing for the accelerated bone formation necessary to accommodate the rapidly growing brain. The development of the rest of the skull occurs through both endochondral ossification (mandible, cranial base), and a combination of the two (temporal, sphenoid, occipital). The spine also has a complex development through a combination of intramembranous and endochondral ossification. Humphrey (1998) demonstrated that in the growing skeleton, the cranium is earliest to reach adult proportions, and starts with an increase in frontal breadth, ending in the mandible and mastoid process. Long-bone diameters reach adult proportions last, with long-bone length being followed by the completion of growth in the pelvis, scapula and clavicle. Later-growing bones are more sexually dimorphic than the earlier-growing elements, with males growing faster and for longer during puberty. This allows for males, who previously lag behind the females, not only to catch up in size but to over-take the females at the end of the adolescent growth spurt (Humphrey, 1998).

4.3   Prenatal and postnatal growth

One of the most important areas of growth research in clinical studies is the assessment of the health, growth and development of low birth weight (LBW) babies, or those who are born small for gestational age (SGA). Low birth weight is associated with children born to mothers suffering from malnutrition, or infectious diseases such as syphilis or leprosy. Children may be shorter at birth due to prematurity, as the result of poor nutritional status in the mother, or due to a twin birth. At 34–36 weeks' gestation, fetal growth generally slows owing to space constraints in the uterus, and in twins, this process occurs when their combined weight reaches that of the singleton fetus at 36 weeks (Falkner *et al.*, 2003). Archaeological studies involving the infant are limited when attempting to identify periods of stress at birth or during the first year of life. In archaeological material, the lack of dental evidence often results in the use of diaphyseal lengths to estimate the age of the fetus, resulting in any SGA children being under-aged, rather than being recognised as premature or sickly when they died (see Owsley and Jantz, 1985). There are some skeletal correlates of SGA that may be of use to bioarchaeologists. Clinical studies indicate that SGA babies have a shorter radius and ulna in comparison to the humerus (Brooke *et al.*, 1984), and that the ratio between the second and fourth digit is reduced in males who are shorter at birth (Ronalds *et al.*, 2002). In addition, in chronic hypoxia brought about by maternal smoking, the fetus will develop both longer upper limbs and reduced length of the tibia (Lampl *et al.*, 2003). Whether such allometric indicators of fetal stress can be translated into dry bone measurements to reflect LBW and SGA in past populations, or whether the signatures of fetal hypoxia would be evident in the polluted environments of medieval London, for example, has yet to be tested.

Lampl and colleagues (Lampl *et al.*, 1992; Lampl, 2005) have suggested that, rather than following a smooth continuous trajectory, infant growth is characterised by short bursts, followed by periods of no growth (2–63 days), and that 90–95% of normal development in infancy is growth free. We do not know if these natural growth rhythms are visible in the skeletal record, and traditionally periods of stasis have been interpreted as representing a stress event in the child's life. King and Ulijaszek's (2000) model of the 'weanling dilemma' dictates that if growth faltering does not occur at 6 months, then successful weaning must have taken place, resulting in nutritional supplementation without significant levels of contamination and subsequent infection (see Chapter 6). In the weanling dilemma model, any initial growth faltering may be hidden by subsequent growth once the child has recovered. If the child dies as the result of weanling diarrhoea, they would be placed in a younger age category and growth faltering would be disguised. Cameron and Demerath (2002) outlined a series of critical or sensitive periods during childhood, in which not only do the correct environmental conditions need to be met to achieve full potential growth, but conditions during these periods will affect later experiences of morbidity and mortality. For example, during fetal growth, soft tissue and organ systems are undergoing rapid transition, and disruption to the normal pattern of growth during this period has been related to susceptibility to cardiovascular disease, diabetes and obesity in adulthood (Barker, 1994).

4.4   Puberty and the growth spurt

The adolescent growth spurt signals the beginning of the development of secondary sexual characteristics, and coincides with the onset of puberty, and the menarche in females. These processes do not exactly correspond. Females begin their increase in the velocity of growth about 1 year (at the age of about 10 years) before they develop the external signs of developing maturity. The male growth spurt occurs at around 12 years of age, at least 6 months after their bodies have begun to develop (Cameron and Demerath, 2002).

That the age of menarche has changed through time has been a matter of debate by historians and growth researchers for many years. It is generally accepted that menarche was common by 14 years in classical antiquity, but began a gradual delay in the 1500s. By the eighteenth century, the age of menarche in Europe was thought to be between 17 and 18 years of age, only returning to 14 years in the modern period (Post, 1971). These fluctuations have been attributed to changing social conditions, nutrition levels and disease, and would have had an impact on the fertility of past populations. The current trend in the last 100 years has been for the age at menarche to decrease by 0.3 years per decade (Eveleth and Tanner, 1990), with wealthier girls consistently achieving menarche earlier than poorer girls. In forensic cases, delayed adolescent growth spurts and a smaller final height attainment can be expected from children undergoing growth hormone treatment for pituitary disorders (Tanaka *et al.*, 2002). Conversely, chemotherapy has been shown to reduce the cartilage growth plate and hence, accelerate epiphyseal fusion (van Leeuwen *et al.*, 2003). Vulnerable adolescents, precisely the ones who enter the forensic record, often spend time on the street, or may be malnourished at their time of death, potentially delaying puberty and maturation. This should be taken into account when estimating the growth and age of these individuals during personal identification. Asymmetry in the epiphyseal fusion has been noted by several authors, with the right side usually showing a fusion stage in advance of the left (Albert and Greene, 1999). Such discrepancies are generally reported as natural variation in skeletal development due to age and sex factors (Helmkamp and Falk, 1990), trauma (see Chapter 8) and biomechanical loading (Roy *et al.*, 1994). However, Albert and Greene (1999) have suggested that the asymmetry found in their early Christian Nubian sample compared to the late Christian sample, suggests nutritional stress in the former. Trauma may also result in asymmetrical fusion in a single bone (see Chapter 8).

In their study of Ladino and Mayan schoolchildren in Guatemala, Bogin and colleagues (1992) hypothesised that boys are less buffered against environmental insults than girls, which is reflected in their final stature outcome, and that the speed and amount of growth during puberty is genetically determined. In her study of growth in a New England private boys' school from 1918 to 1952, Leidy Sievert (2003) showed that boys approaching puberty were susceptible to mumps, measles and rubella due to the amount of energy expenditure on growth. Despite their illness, they became taller than boys who suffered the diseases at an earlier age. Stini (1969) examined the growth status of children from modern Heliconia in Colombia who were under continual stress from protein-deficiency malnutrition. He found that females had the ability to make up this growth in pre-adolescence, but that males were more severely retarded, resulting in less pronounced sexual dimorphism in the stature of the adults. This has potential for the study of stature in adult samples in archaeological populations, where levels of sexual dimorphism may be indicative of childhood and adolescent malnutrition.

In modern children, the average overall pubertal height gain in boys is 28.3 cm compared to 21.0 cm in girls (Tanaka *et al.*, 2002). Although it is tempting to view growth profile deviations in later age groups as being indicative of the puberty growth spurt and onset of menarche, Hägg and Taranger (1982:299) warn that: 'even with adequate records, it may be difficult to locate the pubertal growth spurt before it has passed, since the increase in growth rate is often too small, especially in girls, to be clinically discernible'. One method developed in an attempt to measure growth 'velocity' in archaeological populations has been to estimate diaphyseal length in each age category as a percentage of adult height, attained by combining mean male and female adult stature from each sample (Lovejoy *et al.*, 1990; Wall, 1991). This method has merit when attempting to control for genetic variability in growth within each sample, and perhaps when assessing timing of the puberty growth spurt. Lewis (2002c) compared the percent adult height attainment in children from St Helen-on-the-Walls in later medieval York to those attained by children from Christ Church Spitalfields in post-medieval London (thought to be of French ancestry). At 12 years, the urban medieval children from York had attained 94% of their adult height and remained taller than the London children up to the age of 15 years where the data end. This may suggest that the later medieval children began to experience their growth spurt at 12, while those in post-medieval London had a later spurt, after the age of 15 years.

Interestingly, Hägg and Taranger's (1982) data on Swedish children indicate that fusion of the distal radius occurs only after the adolescent growth spurt, and there is some suggestion that ossification of the iliac crest corresponds to the age of menarche, usually at 12.5–13 years (Scheuer and Black, 2004:331). The study of the onset of puberty in contrasting past populations has the potential to tell us much about secular changes in maturation related to contrasting environmental conditions. Unfortunately, such studies are hindered by a paucity of data. Few children die in the older age categories and until approximately 15 years, it is not possible to determine the sex of skeletal remains with any accuracy. Therefore, developmental studies of older age categories are often based on small sample sizes and are potentially biased by a disproportionate number of males or females in one or other of the samples. For example, comparisons of the growth of individuals from later medieval York and its neighbouring rural settlement showed a discrepancy between the growth profiles after the 12-year-old age category, with height in urban York exceeding that of the rural settlement (Lewis, 1999). It has been suggested that female migration into York was extensive during the medieval period (Goldberg, 1986). Differences between the samples may not simply be showing a delay in maturation, but rather the greater number of females contributing to the mortality sample in York, and their earlier puberty growth spurt. Similarly, studies that aim to examine delayed skeletal maturation in comparison to dental development due to environmental stress will struggle with a lack of children in older age categories when long-bone maturation begins, poor maturational data for when such changes should occur in past populations, and the inability to control for sex.

4.5   Factors affecting growth

Modern studies on the growth of children from different populations and social backgrounds have shown that growth is considerably adaptable, and can be affected by many factors including: nutrition (Metcoff, 1978), infection (Scrimshaw *et al.*, 1959), socio-economic status (Bogin, 1991), urbanisation (Tanner and Eveleth, 1976), migration (Johnston *et al.*, 1975), physical activity (Bogin, 1988b), physiological stress (Eveleth and Tanner, 1990), climate (Panter-Brick, 1997), exposure to lead (Wu *et al.*, 2003), intestinal parasites (Worthen *et al.*, 2001), altitude (Frisancho and Baker, 1970; Delgado-Rodríguez *et al.*, 1998) and even noise pollution (Schell, 1981). A child is most vulnerable to growth retardation during the prenatal period, when maternal age, protein reserves and psychological stress can limit prenatal weight and linear growth (Frisancho *et al.*, 1977, 1985; Mulder *et al.*, 2002). There is also vulnerability in the postnatal period (from birth to 6 years), when the child experiences a series of major biological, psychological and social changes (Haas, 1990). During adolescence the genetic influence on growth is more strongly expressed (Frisancho *et al.*, 1970a, b).

The synergistic relationship between genetics and nutrition in growth attainment is demonstrated by studies of children born to Guatemalan immigrants in California. Guatemalan children are genetically shorter than their Black, Mexican and White peers living in the area but, due to parental economic investment, they are taller than children born in Guatemala (Bogin and Loucky, 1997). In fact, most studies have shown poor growth to be related to poverty and its related factors (e.g. parental education, diet, access to medical resources), making any investigations into environmental circumstances alone extremely problematic if the social status of each individual cannot be determined (Tanner and Eveleth, 1976).

During a stressful period, the growth rate of a child slows until normal nourishment is resumed, with the extent of slowed growth reflecting the severity of the episode and the age at which it occurred (Hewitt *et al.*, 1955; Tanner, 1981). Acheson and Macintyre (1958) demonstrated that in rats, illnesses lasting 48 hours had no effect on final growth outcome, but periods of infection and starvation that lasted for up to 5 days did. As growth is target-seeking, once the child's health has been restored, it has the ability to 'catch up' with its peers and resume growth at an increased rate of up to three times the normal velocity (Tanner, 1981). This ability to reverse retarded growth (homeorrhesis) is impaired with age, and if growth is slowed for too long or too near puberty, increased growth velocity cannot return the child to the optimum level (Prader *et al.*, 1963; Rallison, 1986). Although the exact mechanisms behind catch-up growth are not fully understood, it has been shown that the potential to catch up is greater in older individuals if maturation is also delayed and the growth period is subsequently prolonged. Accelerated growth may advance maturation, shortening the growth period and resulting in stunting (Martorell *et al.*, 1994). Another example of target-seeking growth is demonstrated by the opposite phenomenon of catch-up growth, known as 'catch-down' growth. As Tanner (1986:174) explains: 'during infancy a re-assortment of relative sizes among children comes about: those who are larger at birth grow less, and those who are smaller grow more . . . in a series of studies . . . not only did many small babies catch up to higher centiles, but many large babies sank back to lower ones . . .' Stini (1975) referred to growth retardation as 'developmental adaptation' and argued that this response increased an individual's chance of survival due to fewer nutritional requirements. Seckler (1982), more controversially, suggested that such individuals were in fact 'small but healthy'. Frisancho and colleagues (1973) working in Peru had previously demonstrated that during periods of environmental hardship, smaller women had better success in rearing their children than taller women, perhaps as a result of their improved environmental adaptation. Others argue that since growth retardation has been consistently shown to be correlated with negative health factors in both childhood and later adulthood, short stature should be considered 'acclimatisation' (or short-term adjustment) rather than as evolutionary 'adaptation' to environmental stress (Beaton, 1989; Scrimshaw and Young, 1989).

Studies of the differences in height between children in towns and villages have been carried out since the 1870s. For much of this period the height of rural children surpassed that of their urban peers (Meredith, 1982). For example, Schmidt (1892, cited in Meredith, 1982) studied the growth of German children and found that the urban group were on average 1.8 cm shorter than their rural peers. In London, this difference was particularly marked during the eighteenth and nineteenth centuries, when the Marine Society reported retarded growth of poor children both before and after puberty, and suggested that they must have been small at birth (Floud and Wachter, 1982). Male children from the urban slums were up to 20 cm shorter than children from the urban upper classes (Schell, 1998). By the 1950s this situation had changed, and in Meredith's survey, urban children between the ages of 7 and 10 years exceeded the height of rural children of the same age by 2.5 cm, and in weight by 1.1 kg. This reversal in the trend is often attributed to improvements in sanitation, nutrition and healthcare in the cities after the 1900s (Bogin, 1988a). Today, children living in the urban slums of the developing world are still smaller than their rural counterparts (Bogin, 1988b), whereas in modern industrial countries, differences in growth between urban and rural areas are barely discernible (Schell, 1998). Parental education and income have a more serious effect on growth than whether the child comes from an urbanised environment (Crooks, 1999).

Studies of homeless boys in Nepal indicate that they have better growth and overall health than rural boys of similar age, and of children living with their parents in the urban slum areas (Panter-Brick *et al.*, 1996; Worthman and Panter-Brick, 1996). The poor growth of rural children was related to their limited nutritional intake and greater energy expenditure. While the homeless boys had a degree of independence that allowed them to earn their own money and buy their own food, urban slum boys gave most of their earnings to the family (Todd *et al.*, 1996). In addition to these economic considerations, Tanner and Eveleth (1976) have argued that individuals living in the urban environment are genetically more heterogeneous than those in rural areas, and that this genetic interchange may affect future patterns of growth. Bogin (1991) and Dettwyler (1992) agree that variability in growth within a population may not reflect the degree of environmental adversity, but rather the level of genetic diversity.

4.6   Growth studies: methods and concerns

When attempting to assess the health status of non-adults, growth data coupled with evidence of disease and nutritional stress can provide useful information on the success of adaptation to certain cultural or ecological circumstances. Many standards exist for the growth of modern children from all around the world (Table 4.1) and today, child health studies rely on anthropometric measurements such as height-for-age, weight-for-height and arm circumference (World Health Organisation, 1986; Ulijaszek *et al.*, 1998). Growth researchers assess the variability of an individual's height-for-age in comparison to a reference population, over a long period of time (longitudinal data). In archaeological samples, growth data is obtained by one-off measurements of the diaphyseal length (cross-sectional data), before the fusion of the growing ends of the bones, marking the cessation of growth. However, due to the nature of our data, studies of deceased individuals do not provide true growth curves, nor can they provide information on a child's growth velocity. Rather, we compare the growth profiles of individuals who died within each age category to a modern healthy population, or with the growth profiles of a contrasting archaeological group.

Wood and colleagues (1992) did not see fluctuations in growth patterns in past populations as reflecting anything but periods of high and low mortality. They argued that when the mortality rates are high, both tall and short people enter the archaeological record, providing greater mean stature for that population, but during periods of low mortality, only the frail die. As these individuals are usually shorter, they misrepresent the survivors as a short population. However, information on the age at death of individuals with short stature should identify the 'frail' in a given population, as they will have entered the mortuary record at a younger age (Goodman, 1993). Saunders and Hoppa (1993) revisited these issues and warned that, as short stature is associated with childhood mortality in modern populations, there will be a 'mortality bias' in the samples of children entering the skeletal record. That is, that 'non-survivors' will be short as a result of their disease experience that led to their early death and cannot be compared to modern standards, or be considered representative of the living population from which they were derived. Sundick (1978) argued that the majority of deaths in childhood would have resulted from acute illnesses that would have killed the children too quickly to affect their growth. The problem is that children who are susceptible to acute illnesses have usually suffered previous stress and therefore are already likely to be shorter than their healthier peers who survived into adulthood (Humphrey, 2000). Such a scenario was demonstrated in a study of Nepali boys, where continued and underlying low levels of acute and chronic infection resulted in stunting (Panter-Brick *et al.*, 2001). This study also introduced the caveat for modern comparative data that in rural areas the children rarely reported illness due to their routine experience of ill health, and low expectations of receiving any treatment.

As modern studies have shown secular improvements in growth over time due to the influence of improved living conditions (Bogin, 1988a), direct comparisons of archaeological growth profiles with modern groups to identify retarded growth are inappropriate. Instead, data should be used as a guide to optimal

Table 4.1 *Published modern growth data for children from different geographical regions*

Population	Source	Sample	Notes
American	Anderson and Green (1948); Anderson *et al.* (1964)	134	
American	Hamill *et al.* (1977)	Not given	Boys and girls aged 0–18 years
Belgium	Beunen *et al.* (1988)	4278 boys	Semilongitudinal data for Flemish boys aged 6–19 years
Croatian	Buzina (1976)	87	Boys and girls aged 1–18 from urban and industrial areas
Czech	Blaha (1986)	778	Boys and girls aged 6–18 years
Danish	Andersen *et al.* (1974, 1982)	c.850	Girls and boys aged 0–6 years
Dutch	Roede and van Wieringen (1985)	c.300	Boys and girls aged 0–19 years
East African (Bantu)	Davies *et al.* (1974)	419	Includes weight, skinfold thickness and arm circumference for boys and girls
English	Rona and Altman (1977)	Not given	Boys and girls aged 11 years
Flemish	Roelants and Hauspie (2004)	16 096	For use on children from Northern Europe
Gambian	Billewicz and McGregor (1982)	238	Boys and girls aged 0–25 years
German	Reinken *et al.* (1980)	1420	Boys and girls aged 1.5–16 years of mixed socio-economic status
Hungarian	Eiben and Panto (1986)	3293	Census data for boys and girls aged 3–16 years
Indian	Agarwal *et al.* (1992)	22 850	Affluent Indian children aged 5–18 years
Iranian	Amirhakimi (1974)	131	Compared high and low status, urban and rural growth patterns of girls and boys
Irish	Hoey *et al.* (1987)	c.200	Boys and girls aged 5–19 years from urban and rural environments
Italian	Capucci *et al.* (1982–3)	5500	Boys and girls aged 4–15 years

Population	Reference	Sample size	Description
Japanese	Greulich (1976)	1 391 645	Boys and girls aged 6–20 years
Kenyan	Stephenson et al. (1983)	190	Rural children aged 2–18 years
Norwegian	Waaler (1983)	838	Boys and girls aged 3–17 years
Pakistani	Kelly et al. (1997)	785	Boys and girls aged 5–14 years
Polish	Kurniewicz-Witczakowa et al. (1983)	c.222	Boys and girls aged 0–18 years
Russian (USSR)	Fox et al. (1981)	c.4000	Boys and girls aged 1–4 years
Rwandan	Heintz (1963); Hiernaux (1964)	c.100	Tutsi and Hutu data separate, ages 6–17 years
Spanish	Sandin-Dominguez (1988)	c.400	Boys and girls aged 6–15 years
Swedish	Karlberg et al. (1976)	212	Boys and girls aged 0–16 years from urban environment
Swiss	Prader and Budlinger (1977)	215	Boys and girls from Zurich aged 0–12 years
Various	Maresh and Deming (1939)	80	Infant long-bone lengths measured from radiographs, combined males and females
Various	Maresh (1955)	3200	Follows from Maresh and Deming (1939) to include children up to 18 years; male and female data and epiphyseal length data available
Various	Meredith (1982)	c.896 000	Comparison of urban and rural difference in height for boys and girls in periods 1870–1915 and 1950–1980
Various	Gindhart (1973)	3150 boys 2563 girls	Tibia and radius data for males and females aged 1 month to 18 years

Source: Adapted from Eveleth and Tanner (1990).

growth. Today, growth data are usually presented as percentiles, and the most useful studies provide the 10th, 50th and 90th percentiles for comparison with archaeological data (Humphrey, 2000). Problems with the use of modern data sets for comparison with archaeological material include the fact that long-bones may be damaged post-mortem reducing the available sample size, and that they have different rates of maturation. Records that include the fused epiphysis in modern children are inappropriate for comparison with archaeological samples where the epiphyses remain unfused. This problem can, in part, be compensated for if the modern data set also provides an overlap of measurements for children with both fused and unfused diaphyses, such as Maresh (1955). Here, the percentage contribution of the epiphyses to bone length can be estimated and subtracted from the modern data if necessary.

Powers (1980) tried to remedy the problem of our reliance on modern standards for growth, at least for post-medieval groups, by publishing growth data from all known-age children from archaeological assemblages. Comparisons between different mortuary samples assume that the level of mortality bias is the same. Furthermore, most archaeological samples are multiperiod with cemetery use spanning up to 500 years. Any secular trends in such a population will be hidden by the mean measurements, even though it is these trends that are of interest to the bioarchaeologist. For instance, Albanese (2002) warned against combining data from several sources, even when they are thought to represent those of similar racial groups. His study of femoral lengths of African populations showed a secular change, evident in the combined sample, that did not exist when each group was analysed separately. It should also be remembered that Black populations are consistently smaller at birth, but larger between the ages of 2 and 12 years when compared to White and Asian populations (Garn and Clark, 1976), a factor that is impossible to take into account when comparing the growth of children whose ancestry is so difficult to determine.

Powers (1980) reported that the individuals included in her growth study from post-medieval Britain lagged up to 2 years behind modern children, and up to 6 months behind in their dental development. In fact, dental standards on which our age estimates are based have been shown to be in error of true age by at least 2 years (Lampl and Johnston, 1996), and hence, will under- or overestimate growth achievement when compared to modern samples. Saunders (2000) criticises the variety of dental standards used to age non-adults, which make growth profiles impossible to compare. Variation in the growth patterns of two child samples may simply be a reflection of the different standards used, rather than indicating any environmental diversity. This problem was first highlighted by Merchant and Ubelaker (1977) who produced very different growth curves from the same sample, simply by using different dental ageing criteria. The majority of researchers advocate the use of the Moorrees *et al.* (1963a, b) dental standards for constructing growth curves.

Goode and colleagues (1993) attempted to solve the perpetual problem of small samples in archaeological growth studies. They proposed standardising data to allow any bone to be plotted on the same graph, by expressing the maximum length of all available bones in one individual as a proportion of a given standard (δl_i). This allowed for any individual with a dental age and an intact long bone to be represented in the growth profile. Others have cautioned that in comparative studies, δl_{mean} may actually reflect genetic differences in limb proportions, rather than superior growth (Hoppa and Saunders, 1994; Sciulli, 1994). As with all such studies, this method still relies on the number of individuals for whom dental ages are available and the appropriateness of the modern standard used. Sciulli (1994) has also pointed out that long bones will react differently to external stimuli, with the rapidly growing bones of the lower limb showing greater retardation than the upper limbs in periods of stress. Hence, studies seeking to examine the impact of the environment on child growth generally focus on the tibia and femur. Future comparisons of the relative length of the humerus compared to femoral growth (i.e. allometry) could provide an additional dimension to these studies.

Steckel and colleagues (2002) when faced with the problem of comparing growth data from various groups in the Western hemisphere, expressed growth as a percentage of the age-specific standards provided by Maresh (1955). In each age category, those attaining or exceeding the femoral lengths of the modern standard were classified as having 100% growth, while those falling below the third standard deviation scored 0% growth. Skeletal growth studies need to be carried out on large samples in order to minimise the impact of individual variation, where certain age categories may contain both early and late developers.

Before the choice of method can even be considered, it is important to ensure the quality of the measurements that are taken. Adams and Byrd (2002) demonstrated that experience was no guarantee of accuracy. In their study of measurements taken by forensic anthropologists of various levels of experience, the most common errors they encountered were failure to 'zero out' callipers, recording the wrong measurements and transposing numbers (e.g. 37 mm instead of 73 mm). They stress that inter-observer error is compounded when the measurements are not taken in exactly the same way as the standard data being used to estimate growth or stature. The precision of measurements taken, or the technical error of measurement (TEM), can be calculated using the following equation: $\text{TEM} = \sqrt{\Sigma D^2 / 2N}$, where D is the difference between measurements, and N

is the number of bones measured. The coefficient of reliability (R) is then estimated to assess the percentage of variance between measurements created by
measurement error, using the equation $R = 1 - [(TEM^2)/(SD^2)]$, where SD is
the total measurement error (Ulijaszek, 1998).

4.7   Interpretations of past growth

Modern studies of growth in developing countries are complicated by numerous variables, including socio-economic status, genetic patterns, migration,
altitude and feeding practices that are difficult to control for and add confusion to the picture. These problems are compounded in the past when most
of these variables are unknown. The use of growth to understand environmental conditions of past populations began in 1962 with Johnston's examination
of the children from Indian Knoll, Illinois. Most of these early studies concentrated on Native American samples and the impact of the transition from
hunter–gathering to settled agriculture. For example, Jantz and Owsley (1984)
examined the remains of three Arikara samples from different cultural environments and found that increased morbidity and a decreasing subsistence
base were associated with poorer growth attainment. However, they did not
record the prevalence of infection or malnutrition to correlate with these changing growth patterns. Comparisons of growth in archaeological samples are
also used to demonstrate the environmental and cultural impact of migration
(Saunders *et al.*, 1993b) or colonialism.

A list of available growth data published for archaeological populations is
provided in Table 4.2. The majority of these studies have shown individuals in archaeological samples to be much smaller than their modern peers.
Humphrey (2000) demonstrated that the femoral diaphyseal lengths of children from post-medieval London fell below those of the peers from modern
Denver (Maresh, 1955), although they demonstrated prolonged femoral growth
after the 2-year-old age category. One exception to this general rule comes from
nineteenth-century Belleview, Ontario (Saunders and Hoppa, 1993). Here, children seemed unimpeded by poor health and showed similar growth profiles to
modern children, except under the age of 2 years and around 12 years of age,
when modern children begin the adolescent growth spurt. The similarity in the
growth profiles of deceased and living children in this sample is surprising, and
seems to suggest that ancient growth profiles are not biased simply because
these children from the archaeological sample died (Saunders and Hoppa,
1993).

Another common approach to understanding the health of children in the
past is to match the peaks and troughs of growth with the prevalence of stress

Table 4.2 *Growth data for past populations by period*

Period	Sample	Source
Pre-Dynastic	HK43, Hierakonpolis, Egypt	Batey (2005)
4250–300 BP	Central Californian series	Wall (1991)
3000 BC	Indian Knoll, Kentucky	Johnston (1962), Sundick (1972)
3000–400 BP	Native Americans (various)	Sciulli and Oberly (2002)
c.2050–1680 BC	Asine and Lerna, Middle Helladic Greece	Ingvarsson-Sundström (2003)
350 BC–AD 1350	Wadi Halfa, Nubia	Mahler (1968)
350 BC–AD 1400	Sudanese Nubia	Armelagos *et al.* (1972)
5th–8th centuries AD	UK and Southwestern Germany	Jakob (2003)
6th–7th centuries AD	Altenerding, Germany	Sundick (1978)
4th century AD	Poundbury Camp, Dorset, UK	Mays (1985), Molleson (1992), Farwell and Molleson (1993)
AD 550–1450	Kulubnarti, Nubia	Hummert (1983), Hummert and Van Gerven (1983)
AD 800–1100	Libben, Late Woodland, Ontario, Canada	Mensforth (1985), Lovejoy *et al.* (1990)
Pre 9th century AD	Eskimo and Aleut	Y'Edynak (1976)
9th century AD	Ancient Slavic	Stroukal and Hanáková (1978)
AD 850–1100	Raunds, UK	Hoppa (1992)
AD 850–1650	Raunds, UK	Ribot and Roberts (1996)
AD 950–1300	Dickson Mounds, Woodland–Mississippian	Lallo (1973)
AD 950–1300	Illinois Valley, Woodland–Mississippian	Cook (1984)
AD 950–1550	Wharram Percy, North Yorkshire, UK	Mays (1995)
AD 1000–1200	K2, South Africa	Steyn and Henneberg (1996)
AD 1200	Lake Woodland, Illinois	Walker (1969)
Medieval	Westerhaus, Sweden	Iregren and Boldsen (1994)
AD 1500–1850	Ensay, Scotland, UK	Miles and Bulman (1994)
AD 1600–1835	Arikara	Jantz and Owsley (1984)
AD 1615	Late Ontario Iroquois	Saunders and Melbye (1990)
AD 1700–1750	Morbridge Site, Arikara	Merchant and Ubelaker (1977)
AD 1729–1859	Christ Church Spitalfields, London	Molleson and Cox (1993)
18th–19th centuries AD	London Crypt sample	Humphrey (1998)
AD 1821–1874	St Thomas' Church, Belleview, Ontario, Canada	Saunders *et al.* (1993a)
19th–20th centuries AD	Milwaukee County Almshouse	Hutchins (1998)
19th century AD	Randwick Children's Asylum, Sydney, Australia	Visser (1998)
AD 1850–1859	Raunds Furnells, Wharram Percy, St Helen-on-the-Walls, Christ Church Spitalfields	Lewis (2002a, b)
AD 1869–1922	Freedman's and Cedar Grove, African-Americans	Davidson *et al.* (2002)
AD 1250–1900	Great Plains Indians	Johansson and Owsley (2002)

indicators on the teeth and skeleton. Mensforth (1985) correlated growth retardation with iron deficiency anaemia in the Libben Late Woodland sample, and attributed it to nutritional stress in the weaning period. Lovejoy and colleagues (1990) correlated a high prevalence of infections in children who showed depressed growth patterns at 3 years old. In Britain, Mays (1985, 1995) tested the association of growth with cortical thickness and with Harris lines in the non-adults from Wharram Percy. Studies examining the impact of childhood stress on growth have also been carried out by Wiggins (1991) on non-adults from St Peter's Church, Barton-on-Humber, and by Ribot and Roberts (1996) who compared the prevalence of stress indicators with growth curves from a later medieval and an Anglo-Saxon population. Interestingly, few studies have shown that stress indicators are significantly correlated to growth retardation. The lack of association between height and multiple stress indicators has been attributed to effective catch-up growth when the stress was removed (Mays, 1995; Byers, 1997).

4.8   Bone density

Cortical bone growth, measured by examining cortical thickness, provides complementary data for long-bone growth and health (Johnston, 1969; Larsen, 1997). Various studies have found an association between retarded growth, indicators of nutritional stress and reduced cortical mass (Huss-Ashmore *et al.*, 1982; Saunders and Melbye, 1990; Mays, 2000b). For example, Hummert (1983) examined cortical thickness, diaphyseal length and histomorphology of early and late Sudanese Nubian non-adults and demonstrated that while long-bone length was maintained, cortical thickness declined in the later groups. This suggests that in periods of nutritional stress, height is maintained at the expense of bone robusticity. However, Ruff (2000) has argued that such interpretations are oversimplistic and based on flawed methods, which traditionally compare total periosteal area, cortical area and medullary area on cross-sections of the tibia or femur. Periosteal, medullary and cortical areas have been shown to follow very different growth trajectories, with cortical thickness being more influenced by mechanical loading than nutritional health (Daly *et al.*, 2004). A re-examination of the Nubian material taking into account these caveats showed that bone mineral content increased normally with age (Van Gerven *et al.*, 1985). In order to truly examine osteopenia as the result of malnutrition, Ruff (2000) suggests that non-weight-bearing bones of the upper limb and joint size in relation to long-bone length ('body mass') also need to be taken into account.

4.9   Estimations of stature

Stature estimates for children are not usually attempted in bioarchaeology, because in life their standing height is derived from the diaphyses and the epiphyses, joined by a cartilage growth plate. Instead, data based on diaphyseal lengths are used to assess the growth of a child to measure the progress of its development compared to its peers. The reconstruction of standing height in children is mainly the concern of forensic anthropologists and palaeontologists. The estimation of stature from stride length and footprints in adults has been developed in forensic science (Jasuja *et al.*, 1997; Ozden *et al.*, 2005), but similar data have as yet to be collected for non-adults.

It is our inability to assess the contribution to length and standing height of the cartilaginous growth plate that hinders estimates of stature in skeletonised remains, as it varies in thickness at different times of the child's development, and between individuals. Early attempts to estimate this thickness were made by Seitz (1923) who examined the relationship of epiphyseal length to stature in the tibia and humerus of White adult males. The proximal epiphysis of the humerus made up between 1.3% and 2.2% of the total bone length and the tibia varied between 2.4% and 3.9%. Notably, Seitz's results showed a lack of correlation between the length of these bones and the length of the proximal epiphyses, and also that the length of the epiphyses varied between left and right bones in the same individuals. If these caveats are taken into account, data do exist that allow for the approximate thickness of the growth cartilage and epiphysis to be calculated. Maresh (1955) presents both diaphyseal and long-bone length data for male and female French Canadian White children. His data overlaps in the 10–12-year age groups for boys and girls revealing the epiphyses account for an average of 2.3% bone length in the humerus for boys and girls, and the femoral epiphyses account for 3.8% of bone length in boys and 4.6% of bone length in girls.

Estimations of fetal and child stature were first attempted by Balthazard and Dervieux (1921) and Olivier (1969), who used the stature measurements to estimate age in White French children. But doubts have been raised about the accuracy of the formula published for the radius (Huxley and Jimenez, 1996). There are known differences in length and maturation at term in new-borns from different ethnic groups (Garn and Clark, 1976; Mathai *et al.*, 2004). Himes and colleagues (1977) provide regression equations to estimate stature in Guatemalan boys and girls between the ages of 1 and 7 years, based on the length of the second metacarpal. The authors warn that these children are significantly shorter than their European peers, and within each age group there was a variation of up to 4 cm in stature. Data also exist for stature of Finnish boys and girls based on both diaphyseal lengths of the long bones (0–15 years) (Telkka *et al.*, 1962) and mid-femoral diameter for fragmentary remains (0–15 years) (Palkama *et al.*, 1965). Error in the stature estimates varies from 2.5 to 7.0 cm in the latter method. These regression equations may be of use for White European children, but they are based on data collected from 1950 to 1960, and it is widely known that there has been a secular increase in stature since that period (Henneberg, 1997). Visser (1998) developed formulae to estimate stature and body weight in the remains of non-adults from a nineteenth-century orphanage from Sydney, Australia using methods derived in palaeontology, and based on clinical data reporting the muscle fat and bone widths of modern children. The most reliable estimates were derived by combining stature measurements (maximum length of the humerus, femur and tibia) with mid-shaft diameters. The study revealed that the children from the orphanage were smaller and lighter than their modern counterparts in similar institutions. This result was concordant with documentary evidence suggesting the children were already suffering from chronic illness and malnutrition when they entered the orphanage.

In 1958, Imrie and Wyburn reconstructed fragmentary femora in the case of an adolescent found on the hillside in West Scotland. They used dental wax and comparisons with the bones of children of 'similar maturity' to estimate the approximate length and thickness of the growth plate (Imrie and Wyburn, 1958). The authors acknowledged that estimation of stature from reconstructed bones would introduce errors, that the formulae were based on adults, and that the proportionality of long-bone length to stature changed throughout the growth period, especially during the puberty growth spurt. However, their attempts led to the final identification of the child and the 11 cm difference between the child's recorded height 12 months previously was explained by continued growth. Such attempts to identify the remains of modern children are a necessary evil in cases where any information can allow closure for friends and family. In a forensic context, the stature information gained from medical records in a suspected victim will probably be different to the height of the child when they went missing, due to their continued growth. In this type of situation, even a makeshift family chart may provide useful information that can be matched to the long-bone estimates. In 1968, Snow (Snow and Luke, 1970) carried out a similar procedure when trying to identify a set of skeletal remains that might have belonged to either of two 6-year-old girls, 'Anna' and 'Barbara', who went missing in Oklahoma in 1967. Although the long bones were fragmentary, Snow estimated the length of the diaphyses and cross-checked those against national standards and information provided by the families that suggested Barbara was tall for her age, and Anna was short for her age. Once Snow was happy that the measurements were in conjunction with expected estimates, he extrapolated the rate of growth from the longitudinal growth data provided by Anna's paediatrician, to account for the last measurement, and her height at the time she went missing. From these data, combined with the age-at-death and associated hair and clothing, Snow was able to determine that the remains belonged to the slightly older child, Barbara. This case was unusual as it involved a process of elimination based on two options, and while stature estimates provided crucial additional evidence for the identification of the remains: 'had the two girls been a few months closer in age or a few inches closer in stature, their remains would probably have been indistinguishable.' (Snow and Luke, 1970:275).

Hoppa and Gruspier (1996) have also tackled the problem of reconstructing lengths for fragmentary diaphyses in archaeological samples. They measured the distal and proximal shaft breadths of the humerus, radius, femur and tibia, comparing them to overall length of the bones, but stressed that their measurements are population-specific and should be created for each sample independently. It is feasible that males will have more robust diaphyses than females during growth, and differences in the size of the bones have been noted (Humphrey, 1998; Coussens *et al.*, 2002). In addition, there are documented sex differences in femur/stature ratio in children between the ages of 8 and 18 years, which in unsexed forensic cases cannot be measured (Feldesman, 1992).

4.10   Summary

Growth studies provide valuable information on the health status of children in past populations, and have indicated how contrasting environments and exposure to stress and disease can affect the growth outcome of these vulnerable members of society. For those who survive into adulthood, the stress resulting in growth retardation will have an impact on the function and general health of that population. Despite proving a popular and informative area of research, growth studies present a number of challenges. By comparing modern growth standards with archaeological samples we are comparing the growth of children who died to that of healthy living children from populations known to have seen an increase in optimal height in recent decades. Whether the growth of the children who entered the mortality record accurately reflects the growth of those who successfully went on to achieve adulthood is open to debate. Comparisons of growth in different archaeological samples need to ensure that the same dental ageing criteria have been used, and that measurements have been taken correctly. In forensic cases, stature estimates are often required to allow for the personal identification of an unknown child. However, fragmentary remains and continual growth often cause this to be one of the less reliable identifiers an anthropologist can provide, and these data are more usefully employed to confirm an identity. In order to limit some of the errors inherent in curves derived from archaeological populations, only comparisons between mortuary samples, from similar genetic backgrounds and using the same ageing techniques, should be carried out when trying to assess environmental impact on growth. Information on environmental conditions, levels of migration and the period of cemetery use should also be obtained where possible.

5   *Difficult births, precarious lives*

> [d]eath is the ultimate indicator of failure to adapt . . . [it] may be fruitfully
> viewed as the end result of an accumulated set of biological, behavioral and
> cultural challenges to the individual.
>
> *Goodman and Armelagos (1989:231)*

5.1   Introduction

The transition of a child from a stable uterine environment to the external
environment, with its variety of pathogens and other stimuli, can be viewed as
the first crisis in a human's life. In the womb, the fetus receives nutrients from its
mother, and is protected from the external environment by the maternal immune
system (Gordon, 1975; Hayward, 1978). The response of the mother to social
and economic conditions indirectly influences the fetus. After birth, the physical
and biological environment has a direct effect on the child itself. For a short
time, there is a degree of continuation in the biological link between mother and
child, as breastmilk supplies the child with the nutrients and passive immunity it
needs to grow and to survive the new breed of pathogens it is exposed to (Hansen
and Winberg, 1972). If the child fails to adapt to its new environment, it will die.
The ability of a given population to provide the biocultural means for a child's
survival once it is born provides an insight into its adaptive success. Hence,
the age at death of children from past societies has the potential to provide
the bioarchaeologist with information on obstetrics, disease, socio-economic
transitions, accidents, infanticide and weaning practices, and acts as a measure
of the successful adaptation of a population to its environment.

5.2   Infant mortality rates

Infant mortality has a profound effect upon the crude death rate of a population,
and consequently is considered a sensitive indicator of overall population 'fitness' (Saunders and Barrans, 1999). Infant mortality is calculated as the number
of infant deaths in one calendar year per 1000 live births (Stockwell, 1993).
Infant survivability has been found to be largely dependent on the physical
environment into which the child is born, in particular, the level of sanitation

81

and availability of healthcare facilities. It is the improvement in these factors that is largely believed to have contributed to the rapid decline of infant mortality in England and Wales during the early twentieth century, described as the 'demographic transition' (Williams and Galley, 1995).

Estimated infant mortality rates from modern pre-industrial populations have been found to vary widely up to approximately 200 per 1000 live births (Schofield and Wrigley, 1979). Infant mortality figures for many past populations are uncertain, but in sixteenth-century England it was estimated that around 27% of children died before the age of 1 year (Orme, 2001). In 1662, John Graunt approximated that in London during non-plague years, 36% of children died under the age of 6 years, and that urban deaths exceeded rural ones (Graunt, 1662). Given the inconsistency with which stillbirths and neonatal deaths were reported in the past, historical documents such as the London Bills of Mortality are likely to be an underestimate. By 1907, the death rate in the UK for children aged between 1 and 5 years had dropped by 33%, although infant mortality remained high and averaged around 149 deaths per 1000 live births (Dyhouse, 1978). At this time, infant mortality was highest in the overcrowded urban and industrial mining towns of the Midlands and the North and, overall, higher in the cities than the countryside (Thompson, 1984). This is in contrast to today's statistics, where mortality rates are highest among children in rural areas (Bogin, 1988b).

Much of the evidence for past urban–rural mortality comes from eighteenth- and nineteenth-century documents concerned with social reforms and combating high levels of infant deaths in the cities. Although the Bills of Mortality were first published in 1532 (McNeill, 1979), few studies have been carried out on earlier populations to assess differences in mortality rates between urban and rural groups. Under-representation of the youngest members of these societies makes infant and child mortality rates difficult to assess. Storey (1988) examined the prevalence of prenatal defects and infant mortality in two samples from Mesoamerica. At Teotihuacan, a low-status densely populated urban centre, infant mortality and prenatal enamel defects were higher than at the high-status, low-density site of Copan, reflecting the different stress patterns of the mothers in these two contrasting settlements. In the later periods of urbanisation at Teotihuacan (Tlajinga 33), Storey (1986) found an increase in perinatal, infant and child mortality, and attributed this to the worsening effects of urbanisation on the population.

That male perinates are more susceptible to death than females is widely established in the modern medical literature (Ulizzi and Zonta, 2002). For example, Stevenson and colleagues (2000) examined the outcome of LBW infants in a modern hospital series, and demonstrated that males were more likely to be underweight, had higher rates of neonatal mortality and had a greater need for incubation and resuscitation. They were also more likely to develop pulmonary problems and intracranial haemorrhage than females. Reasons for male vulnerability at birth are a matter of ongoing debate, but it may be the result of natural selection where there is early sparing of the segment of the population vital for birthing and nurturing the next generation (Stevenson *et al.*, 2000), or it may be due to genetic diseases that are liked to the X chromosome. In males, who have only one X chromosome, a genetic disease will develop, whereas females, with two X chromosomes, have a greater chance of only carrying the disease.

In their study of several parish records from England (AD 1550–1649), Schofield and Wrigley (1979) estimated that mean infant mortality rates for males was 137 per 1000 live births compared to 120 for female infants. The non-registration of stillbirths and non-baptised infants in seventeenth-century records affects the accuracy of the registers used by historians to estimate birth and death rates. The rise of infant mortality in 1820 may have been the result of earlier baptism which meant that more infants received funeral services and were included in the registers (Krause, 1969). The absence of infant mortality records for earlier periods has meant that skeletal remains from cemetery excavations are the primary source of evidence concerning infant death in the past. This archaeological evidence does, of course, have many limitations, which have been addressed at length in the literature (e.g. Sundick, 1978; Gordon and Buikstra, 1981; Johnston and Zimmer, 1989; Goode *et al.*, 1993). Differential burial practices, together with preservation and recovery biases, limit the extent to which infants excavated from cemetery populations represent the true proportion of those dying within a particular society at any one time (Saunders, 2000; and see Chapter 2). Furthermore, anthropologists are unable to calculate actual mortality 'rates' from this type of information because data concerning the total population at risk (i.e. the number of live births) are not available. It is also frequently impossible to narrow the dates of archaeological samples into yearly, or even decadal, increments.

Despite these shortcomings, Saunders and colleagues (Saunders and Barrans, 1999; Saunders, 2000) have convincingly argued the necessity and advantages of studying infant mortality rates from archaeological cemetery evidence. It is unfortunate that in many studies it is issues of infanticide (Smith and Kahila, 1992; Mays, 1993, 2003; Gowland and Chamberlain, 2000), and problems with under-representation of infants in palaeodemography that are the main focus of attention. In some cases, the youngest children are excluded from studies altogether (Ribot and Roberts, 1996; Paine, 2000; Bocquet-Appel, 2002). This is disappointing. Infant mortality studies have the potential to provide information on the population's ability to adapt to its environment, seasonality of mortality, cultural practices, demography, maternal health, disease epidemics, birthing practices, infant feeding and social attitudes towards children.

5.2.1   Endogenous versus exogenous mortality

Infant mortality rates are traditionally subdivided into those who die before birth (late fetal or stillbirths), those dying at birth or within the first 27 days of extrauterine life (neonatal mortality), and those who die between 28 days and 1 year of age (post-neonatal mortality). It is an unfortunate drawback of studies of perinatal mortality that we are unable to determine whether a child was premature, stillborn or survived for the first few days of life (Saunders and Barrans, 1999). It is generally accepted that, in the past, children under the age of 28 gestational weeks were unlikely to survive the extrauterine environment due to the immaturity of their vital internal organs.

Clinically, neonatal mortality is considered to reflect the endogenous state of the infant as the result of genetic and maternal influences (e.g. congenital anomalies, prematurity, low birth weight, birth trauma), and post-neonatal mortality is seen to reflect the influence of the child's external environment or exogenous factors, for example, infectious diseases, poor nutrition, poisonings, accidents (Wiley and Pike, 1998; Scott and Duncan, 1999). This 'biometric model' developed by Bourgeois-Pichat (1951) works on the premise that all exogenous deaths occur within the first month and that exogenous mortality is proportional to the number of days after the first month that the infant survives. The ability of a given population to provide the biocultural means for a child's survival once it is born provides an insight into its adaptive success. Today in industrialised countries, sanitation measures ensure that the vast majority of infant deaths are due to endogenous factors (Stockwell, 1993), but the reverse was true in urban centres of the past (Vögele, 1994). The biometric model has been used by historical demographers to identify, among other things, breast-feeding patterns. For example, Knodel and Kinter (1977) demonstrated that in areas where breastfeeding was common practice, exogenous mortality remained low until weaning, and where breastfeeding was uncommon, mortality was high in the earlier months of life, perhaps due to exposure to infected feeding vessels, pathogens in food, and the loss of the protective qualities of the breastmilk.

Historical evidence for medieval Europe suggests that neonatal deaths were higher in rural areas than in urban communities, where environmental factors have a greater effect on child mortality, producing higher post-neonatal deaths (Vögele, 1994). This pattern has also been recorded in archaeological material from elsewhere. Saunders and colleagues (1995) estimated the proportions of

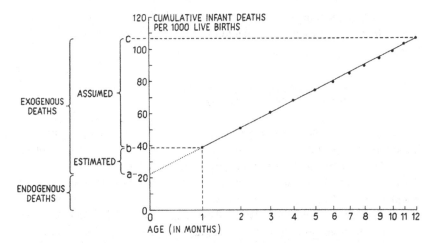

Figure 5.1 Biometric model for estimating neonatal and post-neonatal mortality.
From Knodel and Kintner (1977:392), reproduced with kind permission from the
Population Association of America.

neonatal to post-neonatal deaths in infants from Belleview, Ontario. They found
that 74% of infant skeletons were post-neonates compared to 26% neonates,
reflecting the stronger exogenous influence of infant death in this industrialised
Canadian sample. In a study of historical records from Penrith in Cumbria, UK
(AD 1557–1806), Scott and Duncan (1998) found that both neonatal and post-
neonatal mortality fluctuated in accordance with the rise and fall of wheat prices.
Neonatal mortality lagged behind by 1 year, suggesting that early mortality
was indirectly related to the nutritional status of the mother during pregnancy,
whereas post-neonatal mortality was directly related to lactation and weaning
during the first year of life.

In historical demographic studies, cumulative exogenous mortality is proportional to the $\log^3 (n + 1)$ where n is the age in days within the first year
of life. This cumulative death rate is then plotted on a graph, with the vertical axis representing the number of deaths, and the horizontal axis the days
in a year, spaced according to the above formula. The points representing
the cumulative total of deaths will lie on a straight line or very close to it
(Fig. 5.1). Projecting a vertical line to the left until it cuts the vertical axis enables
exogenous and endogenous deaths to be separated (Knodel and Kintner, 1977;
Wrigley, 1977). For archaeological populations, we do not know the number
of infant deaths per year because of the long time-span of our cemeteries, nor
can we age our skeletons as accurately as days; therefore, such statistics cannot be applied. A crude estimate of exogenous and endogenous deaths can be

established by looking at the numbers of individuals dying between the ages of 38–40 weeks of gestation compared to 41–48 weeks (Saunders *et al.*, 1995; Lewis, 2002c).

5.3   Reconstructing child mortality

Once the crucial infant period has passed, studies of child mortality can provide information on weaning age, exposure to infectious diseases, and susceptibility to accidental trauma. After 1 year of age, children are generally at risk of death during weaning, which usually occurs around the age of 2 years in pre-industrial populations (see Chapter 6). They are also at risk in later years when diseases that may have been contained in early childhood (e.g. tuberculosis) can re-emerge, and when the transition to adulthood can leave the child vulnerable to trauma through warfare and occupation. It is widely accepted that children represent the most demographically variable and sensitive section of the human life cycle (Roth, 1992). In palaeodemographical studies, fears of under-representation have led some researchers to omit analysis of individuals under the age of 10 years from their studies (Buikstra and Konigsberg, 1985). Brothwell (1986–7) has suggested that in developing countries deaths between 1 and 4 years of age are greater or equal to infant deaths due to infectious disease such as whooping cough, measles and pneumonia. Examination of the English medical literature from the eighteenth and nineteenth centuries shows that children under 10 years were particularly susceptible to accidental deaths due to burns, drowning and poisoning (Forbes, 1986).

Schofield and Wrigley (1979) estimated that in England between 1550 and 1649, the mean mortality rate for children between 1 and 4 years was 83 and 80 for males and females respectively, and between the ages of 5 and 9 years fell to 38 and 35 per 1000 live births. The very low death rates are probably the result of under-registration of infant deaths; however, the improved survival rates for older children is still of note. Table 5.1 summarises child mortality patterns derived from archaeological skeletal collections in the published literature, divided into the most convenient age categories dictated by the data. Of the 42 samples (9658 individuals) included in this review, the majority showed a higher percentage of infants (29.5%) and children in the 1–4-year age category (27.5%) than in the age categories of 5–9 years (16%) and 10–19 years (20%). There are a few exceptions to this pattern. The under-representation of infants in the Roman and Anglo-Saxon cemeteries is striking, and it is clear why many authors have suggested alternative burial areas as a cause for this discrepancy (see Chapter 2). The greater number of individuals aged between 10 and 19 years may also reflect the inclusion of children when they reached 'adulthood' around the age of 12 years (Crawford, 1993). Child mortality profiles of children from later medieval England, and in particular York, tend to peak between 5 and 9 years, for example, at Wharram Percy, St Helen-on-the-Walls, Jewbury and Fishergate. Lewis (2002a) found that children who survived into the next age category (6.6–10.5 years) at Wharram Percy and St Helen's had a higher prevalence of sinusitis, cranial trauma and infection, suggesting an increase in their exposure to environmental stress. In the later medieval period many children from the age of 7 were put to work as apprentices (Cunningham, 1995) and records of this practice have been identified as early as the thirteenth century in England (Dunlop, 1912). Interestingly, this peak in the older children does not appear at Westerhus in Sweden (Brothwell, 1986–7), or in the age sample of non-adults from Hungary (Acsádi and Nemeskéri, 1970). The greater number of children entering the mortuary sample in these age groups in the Yorkshire samples may be related to the hazards they encountered when they began employment, and the exposure of the urban immigrants to new environmental pathogens.

In other periods and places, Clarke (1977) reported a peak in non-adult mortality at 5 years at Dickson Mounds, Illinois and three other samples from Sudanese Nubia. Whereas the majority of children recovered from the Indian Knoll in Kentucky (Johnston and Snow, 1961) and at Teotihuacan in Mexico (Storey, 1986) were infants. This contrasts with the lack of children under 1 year of age recovered from the K2 and Mapungubwe sites in South Africa (Henneberg and Steyn, 1994). The analysis of child mortality from contrasting sites and periods can provide information on the activities of children and their exposure to risk and disease at different ages. Their inclusion in cemeteries and the mode of burial can also provide information on the age at which these children become accepted members of the community. As the behaviour of children is dictated by society, wherever possible, such analyses should be informed by cultural data.

5.4   Infanticide

[t]he feeble wail of murdered childhood in its agony assails our ears at every turn, and is borne on every breeze . . . In the quiet of the bedroom we raise the box-lid, and the skeletons are there . . . By the canal side, or in the water, we find the dead child. In the solitude of the wood we are horrified by the ghastly sight; and if we betake ourselves to the rapid rail in order to escape the pollution, we find at our journey's end that the mouldering remains of a murdered innocent have been our travelling companion; and that the odour from that unsuspected parcel truly indicates what may be found within.

William Burke Ryan (1862:46)

Table 5.1 *Child mortality data for a variety of archaeological sites*

Period	Site	Number of non-adults	Percentage non-adults in each age group				Reference
			Infants	1–4	5–9	10–19	
Palaeolithic	Taforalt, Hungary	104	41	37.5	10	11.5	Acsádi and Nemeskéri (1970)
Neolithic	Khirokita, Cyprus	46	74	15	6.5	4	Angel (1969)
Copper Age	Tiszapolgár-Basatanya, Hungary	43	5	19	28	49	Acsádi and Nemeskéri (1970)
Bronze Age	Lerna, Greece	140	60	0.7	7	11	Angel (1971)
Iron Age	Danebury, Hampshire, UK	50	28	40	34	34	Hooper (1991)
Pre-Dynastic	Naga-ed-Der, Egypt	123	15	28	25	32	Lythgoe (1965)
Dynastic	Gebelan and Asuit, Egypt	166	4	25	31	40	Masali and Chiarelli (1969)
4800–3700 BC	Carrier Mills, Illinois, USA	62	69	6.4	5	19	Bassett (1982)
3000 BC	Indian Knoll, Kentucky, USA	420	40	0.2	13	27	Johnston and Snow (1961)
150 BC–AD 750	Teotihuacan, Mexico	111	53	13.5	17	17	Storey (1986)
100 BC–AD 400	Gibson-Klunk, Illinois, USA	191	29	36	15	20	Buikstra (1976)
Classic Period	Greece	70	56	1	13	10	Angel (1969)
Romano-British	Poundbury Camp, Dorset, UK	374	43.5	21	14	20.5	Farwell and Molleson (1993)
Romano-British	Lankhills, Hampshire, UK	118	27	29	27	17	Clarke (1977)
Roman	Intercisa and Brigetio, Hungary	107	2	11	13	30	Acsádi and Nemeskéri (1970)
Roman	Kaszthely-Dobogó, Hungary	37	13.5	43	24	19	Acsádi and Nemeskéri (1970)
Romano-British	Brougham, Cumbria, UK	41	0	54	29	17	Cool (2004)
Romano-British	Cannington, Somerset, UK	148	15	51	10	24	Brothwell et al. (2000)
Anglo-Saxon	Beckford and Worcester, UK	38	0	10.5	34	55	Wells (1996)
Anglo-Saxon	Raunds Furnells, Northamptonshire, UK	208	32	20	15	33	Powell (1996)

Period	Site	N					Reference
Anglo-Saxon	Edix Hill, Cambridgeshire, UK	48	2	21	33	44	Duhig (1998)
Anglo-Saxon	Empingham II, Rutland, UK	62	0	16	27	56	Timby (1996)
Anglo-Saxon	Castledyke South, Lincolnshire, UK	44	0	29.5	16	54.5	Boylston et al. (1998)
9th century AD	Sopronkőhida, Hungary	72	29	35	22	14	Acsádi and Nemeskéri (1970)
AD 970–1200	K2 and Mapungubwe, South Africa	85	0	56	21	20	Henneberg and Steyn (1994)
AD 1050–1150	Meinarti, Sudan	101	10	58	17	15	Swedlund and Armelagos (1969, 1976)
AD 1300–1425	Arroyo Hondo, Mexico	67	43	30	7	19	Palkovich (1980)
13th century AD	Kane, Illinois, USA	60	28	20	18	33	Milner (1982)
Later Medieval	St Helen-on-the-Walls, York, UK	200	6	22.5	56.5	12.5	Lewis (2002c)
Later Medieval	Wharram Percy, Yorkshire, UK	303	26	23	43	8	Lewis (2002c)
Later Medieval	Westerhus, Sweden	225	50	22	11.5	16	Brothwell (1986–7)
Later Medieval	Jewbury, York, UK	150	8	27	32	33	Lilley et al. (1994)
Later Medieval	St Andrew, Fishergate, York, UK	89	12	23.5	26	38	Stroud and Kemp (1993)
Later Medieval	St Nicholas Shambles, York, UK	51	6	31	20	43	White (1988)
Later Medieval	St Gregory's Priory, Canterbury, Kent, UK	22	18	45	32	4.5	Anderson (2001)
Later Medieval	Hungary (all)	4360	46	25	11	19	Acsádi and Nemeskéri (1970)
Post-Medieval	Christ Church Spitalfields, London, UK	186	31	44	19	6	Lewis (2002c)
Post-Medieval	St Bride's Church, London, UK	21	0	67	19	14	Scheuer (1998)
Post-Medieval	Bathford, Somerset, UK	63	6	35	35	24	Start and Kirk (1998)
AD 1500–1900	Ensay, Scotland, UK	200	0	79.5	15	5.5	Miles (1989)
AD 1750–1785	Larson, South Dakota, USA	441	57.5	21	11	10	Owsley and Bass (1979)
AD 1800–1832	Leavenworth, South Dakota, USA	211	15	59	11	15	Bass et al. (1971)
Total		9658	29.5	27.5	16	20	

Infanticide has a long history. The deliberate killing of newborns was practised in many diverse societies from hunter–gatherer tribes including the Yanomamo Indians, the African !Kung San, the Australian Aborigines and the Inuit, to the industrialised nations of Imperial China and Japan (Tooley, 1983). In Western Europe, infanticide is acknowledged from the Roman to the Victorian periods. In modern human societies that overtly practise infanticide, children have been found to be at risk if they are illegitimate, deformed, born too close in time to another sibling, or are of the wrong sex in a society that values one over the other; usually males over females (Scrimshaw, 1984). Hrdy (1994) argues that infanticide is the most extreme form of 'fitness trade-offs' for a family, when abandonment, wet-nursing or fostering is not an option. Today, the legal definition of infanticide varies between countries, as in some societies, infants are not recognised as fully human and are unnamed for periods that range from 3 months to 3 years (Tooley, 1983). However, the medical concept of the practice remains the same. Infanticide refers to the deliberate killing of a newborn infant by its mother, usually within minutes or hours after birth (Knight, 1996).

Historical evidence for infanticide indicates that in ancient Greece and Rome, the practice of child abandonment was widely practised. With Christianity came an emphasis on the sanctity of life and chastity, and in the eleventh century, an accusation of infanticide would result in the excommunication of the mother for life, or her removal to a nunnery (Hanawalt, 1993). In later periods women began to migrate to urban areas as domestic servants. The fear of social 'ruin' due to an illegitimate child led to infanticide as a safer means of disposing of an unwanted pregnancy than abortive or surgical intervention, which might lead to the death of the mother (Sauer, 1978). In the 1500s women caught committing infanticide risked punishment by being buried alive, drowned or decapitated (Langer, 1974), something that perhaps should be considered when excavating deviant burials containing females from this period. By the 1850s, newspapers reported that infanticide was rife in London and the surrounding areas, and painted vivid images of streets littered with the bodies of dead babies (Ryan, 1862). During this period, foundling hospitals and 'baby-farms' were considered the worst culprits of infanticide. This led to their regulation in 1872 under the Infant Life Protection Act and the foundation of foundling hospitals in major cities such as Florence, Paris and London (Langer, 1974). By 1880, medical advancements and the introduction of contraceptives marked a decline in the number of abandoned and murdered infants (Sauer, 1978).

Evidence for infanticide in the archaeological record is indirect and has tra-ditionally been inferred from chance finds within living quarters, clusters of infants, their absence in general cemeteries or unusual male-to-female ratios in the adult sample. For example, in Anglo-Saxon Great Chesterford in Essex, six of the 15 infants from the site were found in one grave, and all were under

40 weeks (Crawford, 1999). Archaeological evidence for probable infant sacrifice comes from Springhead Roman Temple in Kent, where four infants were found, one at each corner of a shrine. The burials represented two periods of deposition, with one intact and one decapitated infant being deposited on each occasion (Penn, 1960). Excavations from an earlier site near the Temple revealed 47 infants under a kitchen floor (Watts, 1989). Recent excavations of a well at Athens containing 450 perinates and 130 dogs has prompted numerous interpretations including infanticide, ritual sacrifice and the usual deposition of individuals not yet viewed as members of society (Liston and Papadopoulos, 2004). In pre-Columbian Cuba, numerous adult–infant burials raised questions about possible infanticide after the death of a parent. DNA analysis of a female burial also containing a 4-year-old child showed that they were related, prompting archaeologists to argue that the child was sacrificed when its mother died, perhaps due to some congenital condition (Lleonart *et al.*, 1999). However, neither the mother nor the child had obvious signs of pathology on their skeletons and both may have died during an accident. DNA analysis of all the adult–infant burials would need to be carried out to substantiate such a dramatic hypothesis. The inclusion of a child with an adult dying around the same time to ensure protection during their passage to the afterlife is a much more common practice (Roberts and Cox, 2003).

Infant burials have been associated with foundations and shrines as early as the Iron Age, with infant remains recovered at the shrine at Uley, Gloucestershire, Maiden Castle, Dorset (Watts, 1989) and Roman Winchester (all UK) (Scott, 2001). Wicker (1998) carried out a survey of infant remains from Iron Age sites in Scandinavia where they were found strewn through middens, within the stone packing of cairns, and in wells and bogs. The association of these remains, often with young animals and in bodies of water, may suggest some form of ritual sacrifice. For example, Mays (2000a) outlined the Germanic practice of testing a baby's strength by holding it under running water; if it drowned the body was left in the river. The disposal of infants in rivers, or their exposure to scavengers may well account for the lack of infant remains within pre-medieval cemeteries. The unexpected presence of infant bones mixed in with juvenile animal remains between cairn stones in Scandinavia led to many infant remains being missed and omitted from the initial skeletal assessment. Wicker (1998) comments that in later medieval Scandinavia, infant burials within cemeteries become more common, although she suggests that cases of infants being deposited outside cemetery walls and, in one case, under a clay floor, suggests the infants were unbaptised, and were perhaps victims of infanticide. Scott (1991) prefers a more engendered approach to the evidence. The placing of infant remains within agricultural areas such as ovens and corn driers in the Romano-British villas of Barton Court Farm,

Berkshire and Hambledon, Hampshire led Scott to argue that the presence of infants under domestic dwellings ('in subgrundaria') and within agricultural spaces, represented the women's attempts to lay claim to their domains and greater control over agricultural production in times of female oppression, rather than infanticide.

In Italy, North Africa and Britain, archaeological evidence has illustrated a transition in the treatment of infants through the early and late Roman periods, perhaps with the adoption of Christianity (Watts, 1989). For example, in pre-Roman Punic Carthage, North Africa (750–146 BC) a large, rectangular walled enclosure was discovered next to the settlement, full of urns containing infants, children up to 4 years old and young animals. At the height of the enclosure's use (400–200 BC) it is estimated that 20 000 urns were deposited, or roughly 100 per annum (Lee, 1994). This site has been interpreted as representing a Punic 'tophet' or sacrificial area, where children were sacrificed to Baal Hammon and his consort Tanit (Norman, 2003). A study of the skeletal remains revealed that 95% of the children were infants, compared to 51% in a control sample from a cemetery of a similar period (Avishai and Smith, 2002). Lee (1994) has suggested a gradual transition from the sacrifice of infants and their animal substitutes, to the sacrifice of older children in the fourth century BC. By the Roman period the pattern of infant burial had changed, and although the bodies of very young children were still treated differently from adults, the Roman tradition of welcoming a child into the family within several days of birth made infant sacrifices abhorrent. Instead, neonates and infants are often recovered from amphorae in specialised cemeteries such as those at Yasmina in Carthage (Norman, 2002), and near Lugnano in Teverina, Italy (Soren and Soren, 1999), with older children distinguished from adults in general cemeteries by their location, gravegoods or body position (Norman, 2003).

5.4.1   The bioarchaeology of infanticide

The study of infanticide in biological anthropology relies on the age assessment of fetal remains in cemetery populations. Suggestions of birth spacing by infanticide in the Palaeolithic have been put forward by Divale (1972) and Ehrenberg (1989). The discovery of an infant (3.5–5 months) with rickets buried within a cave in South Africa led Pfeiffer and Crowder (2004) to suggest that, even though overtly ill, the child was not a victim of infanticide. However, they argued that if infanticide had been practised in this period, it would have been carried out in the perinatal period, when the child in question would have appeared normal. Traditionally, anthropologists and historians have considered female infanticide to be the norm, with high ratios of adult males to females in cemeteries often cited as evidence of this (Divale, 1972). In 1993, Mays argued that female infanticide accounted for the preponderance of adult males in Romano-British cemeteries, and that the infanticide was limited to female children. The higher ratio of adult males to females has also been suggested as illustrating sex-biased infanticide in the Iron Age, Roman and early medieval periods where females may prove a financial burden (Divale, 1972; Coleman, 1976; Wicker, 1998; Mays, 2000a). This theory is riddled with problems, as many factors such as migration, warfare, differential burial and susceptibility to disease will all affect adult male-to-female ratios in a cemetery sample.

Arguments against female-dominated infanticide have been provided by Rega (1997), who suggested that the reason for the excess of females dying between the ages of 1 and 6 years in the Mokrin Bronze Age cemetery in Yugoslavia was that there were more of them available to enter the archaeological record. That is to say, males were being selected for infanticide and then buried elsewhere. Sex estimation of these remains was made on the basis of gravegoods, and it may be that male children did not receive a 'male' burial until they were older. Smith and Kahila (1992) reported on the remains of 100 infants in a late Roman–early Byzantine sewer in Ashkelon, Israel. All the infants were neonates and the absence of the neonatal (birth) line on the dentition led the authors to argue that the sewer was used to bury unwanted infants. Further analysis of the DNA from these remains revealed that of the 19 specimens that could be tested, 14 were male and only five were female. It has been suggested that perhaps illegitimate females were reared to continue in the profession of their mothers, as prostitutes of the bathhouse (Faerman *et al.*, 1997, 1998). In 1999, Waldron and colleagues also carried out DNA analysis on infants from the Romano-British villas of Beddingham and Bignor, both in Sussex. Once again, the remains were not exclusively female. In the light of these studies, Mays and Faerman (2001) returned to the Romano-British samples to test the theory that females were predominantly killed in favour of males, and again, of the 13 cases in which sex identification was successful, nine were males and four were females. These studies all suggest that, if these neonates were victims of infanticide, male preference was not practised in Israel or in Britain during the Roman period. As male fetuses and infants are more susceptible to mortality and disease than their female peers (Stevenson *et al.*, 2000), the higher ratios of males within an infant sample would be expected at these sites if they, in fact, represented cases of stillbirths or neonatal deaths from natural causes, due to male vulnerability at this age.

Criticism of the anthropological evidence for infanticide in archaeological populations does not rest with sex identification and small sample sizes. In 1993, Mays examined the ages of infant remains in Roman cemeteries and noted a peak in the number of children dying at term (38–40 weeks), compared to later medieval Wharram Percy, where infant mortality displayed the normal 'flat' distribution that included younger and older infant deaths. This, Mays (1993) argued, suggested Roman children were being killed at birth. Recent anthropological studies have begun to question the use of regression equations to derive age from long-bone length. A number of authors have demonstrated that ageing techniques based on regression equations are statistically biased towards producing an age structure akin to that of their reference population (e.g. Bocquet-Appel and Masset, 1982, 1996; Konigsberg and Frankenberg, 1992; Lucy *et al.*, 1996; Aykroyd *et al.*, 1997, 1999; Konigsberg *et al.*, 1997). Numerous studies have since attempted to rectify this problem using Bayesian statistics. Bayesian data analysis allows us to make inferences from data using probability models for observable quantities (known-age data) and for quantities that are unknown (archaeological data), but that we wish to learn about (Gelman *et al.*, 1995). In their reassessment of infanticide in Romano-British samples using Bayesian statistics, Gowland and Chamberlain (2002) demonstrated that infant mortality profiles actually fell within the parameters expected if fetal stillbirths and neonatal deaths were included within the burial sample. At present, there is no unequivocal evidence for infanticide in British archaeological material.

5.4.2   Infanticide and the law

The law regarding the murder of infants has a long history. In twelfth-century Europe, all child killing was considered murder and women were presumed guilty and were sentenced to death, unless they could prove otherwise (Sauer, 1978). The leniency of juries against such a harsh penalty in infanticide cases, both in the UK and colonial USA (Scott, 1999:78), finally led to a change in the law in 1803, when the lesser offence of 'concealment of birth' was introduced, punishable by 2 years' imprisonment (Behlmer, 1979). Infant murder was not a separate legal entity in English law until the establishment of the Infanticide Acts of 1922 and 1938, and the statutory registration of stillbirths in 1927 (Hart, 1998). The Infanticide Act allowed that the mother may have been driven to murder as a result of extreme stress caused by giving birth or lactation. Today, in both the USA and the UK, it is the presumption of the court that the child was stillborn, and the onus is on the prosecution to prove that the child had a separate existence once completely expelled from its mother. In 1929, the Infant Life Preservation Act set 28 weeks' gestation as the period at which a child could be considered viable, but with modern medical intervention this period has been reduced to 24 weeks (Knight, 1996).

For forensic pathologists, proof of infanticide is one of the most difficult aspects of their work. Often, the lesser charge of 'concealment of birth' is offered when a child's body is found abandoned under newspapers, plastic bags, in rubbish tips, toilets or under blankets. Such modes of disposal traditionally used for the disposal of waste, as an immediate and convenient mode of concealment, suggest the death was unplanned (Winnik and Horovitz, 1961–2). With advanced decomposition, the pathologist may call on the forensic anthropologist to assist in the analysis, and this task is greatly complicated by the lack of any of the soft-tissue evidence, such as the irregular cutting of the umbilical cord, extent of soft-tissue decomposition, cellular autolysis or, in the eyes, petechiae of strangulation. Strangulation is particularly challenging for the anthropologist as fractures of the hyoid, identified in adult skeletons, will not be evident in an infant where this bone is still largely cartilaginous. Evidence of anencephaly and spina bifida in fetal remains will suggest that, despite the term of the child, its chances of life were limited. As in the past, children are most at risk when they are born to unmarried mothers, with asphyxiation the most common mode of killing, with closed head trauma running a close second (Mendlewicz *et al.*, 1999). In 1873, the *Lancet* (cited in Knight, 1996) reported on a unique mode of killing whereby the midwife or mother would stick a needle under the top lid and pierce the orbital plate, but such subtle evidence is unlikely to be found in skeletal material. Evidence for head trauma or stabbing in the chest will also be difficult to identify in thin and fragile infant remains. Nevertheless, diatoms in the marrow of the femoral shaft have been recovered from skeletonised remains of children who were drowned in open water shortly after their birth. In forensic cases, the presence or absence of the neonatal line is used to assess whether a child was stillborn, or was possibly the victim of infanticide (see Chapter 7). Visible as a microscopic hypoplastic defect on the deciduous teeth, the neonatal line is thought to represent a period of enamel stasis after the stress of birth. Research on the survivability, location and visualisation of the neonatal line has raised serious questions about the reliability of this feature (Sarnat and Schour, 1941). Given the rate of enamel deposition is 4–4.5 μm per day, it is also important to note that the neonatal line may only be visualised if the child has survived for up to 7 days after birth when the stress of being born has subsided and the enamel resumes its mineralisation (Whittaker and MacDonald, 1989). Due to this delay in the appearance of the line, caution is advised in forensic cases where its absence is used to suggest a stillbirth to rule out the possibility of infanticide. Nevertheless, Humphrey and colleagues (2004) have demonstrated that a marked drop in strontium : calcium ratios between the last prenatal enamel deposits and first postnatal deposits can aid in identifying the neonatal line in cases where the microscopic features are unclear.

5.5   Summary

When using data derived from cemetery samples it is important to remember that we are actually measuring *burial rates* and not mortality rates, as cultural practices may dictate whether and where certain individuals were placed within a cemetery. The study of infant mortality in past populations can provide us with valuable information on the success of a population to adapt to its environment and protect the most vulnerable members of its society. Estimates of neonatal versus post-neonatal mortality may help us untangle some of the causes of these deaths, although we can age our infants less accurately than in clinical studies. Past child mortality rates are generally less well studied, but the age at which these older children died, when coupled with evidence of gravegoods, pathology and trauma, can provide us with a glimpse of their activities during daily life, and when they might have become fully fledged adults. Perhaps the most widely explored and controversial aspect of mortality studies in past populations is the issue of infanticide. Previously, clusters of infants in non-cemetery sites or hidden under the floors of houses, in addition to their complete absence from some burial grounds, was enough to convince archaeologists that infanticide was being practised. Anthropological analysis using DNA, the neonatal line and a re-examination of our statistics has yet to produce unequivocal evidence for this practice.

6   Little waifs: weaning and
dietary stress

6.1   Introduction

Since the early 1980s, a combination of osteological indicators have been used
to provide indirect evidence for malnutrition in past populations. Non-specific
metabolic stress is evident with the presence of dental enamel hypoplasias,
radiographic Harris lines in the long bones and porotic lesions of the skull
known as cribra orbitalia and porotic hyperostosis. These lesions can only form
in the developing child, and with the exception of dental enamel hypoplasias,
will all disappear as the child recovers. Less commonly reported are the pathological lesions associated with specific malnutrition, including rickets (vitamin
D deficiency) and infantile scurvy (vitamin C deficiency). Just as today, a malnourished child in the past was unlikely to have been deficient in one single
dietary element, and we should expect a combination of lesions associated with
a lack of iron, zinc, calcium, protein and a multitude of vitamins, including
some lesions that may indicate deficiencies we have yet to identify.

The frequency and age at which these specific and non-specific metabolic
lesions occur are often argued to coincide with weaning. Breastfeeding and
weaning practices have been studied in depth by nutritionalists, biological
anthropologists, palaeopathologists and, more recently, bone chemists. These
studies have examined samples from the Palaeolithic, when the 'hominid
blueprint' for weaning duration was thought to be established, to the introduction of artificial formulas in late-seventeenth-century Europe. Today, children
are weaned at 1–5 years depending on the culture. In modern industrial societies,
medical practitioners advocate the introduction of solid foods after 6 months,
with a total cessation of breastmilk around 12 months of age (Dettwyler and
Fishman, 1992). This chapter will explore the advantages and limitations of the
methods and lesions used to examine the age at weaning, weaning diet and the
nutritional health of non-adults in the past.

6.2   Properties of human breastmilk

Infants have high nutritional requirements due to their rapid growth and a limited
tolerance for dietary deficiencies (Chierici and Vigi, 1991). During gestation,

97

the growing child will accumulate stores of essential nutrients such as calcium, vitamin D, vitamin C, iron and zinc via placental transfer from the mother. These stores will sustain them in the first few months of life. Nutritional deficiencies in the mother will limit the supply to the fetus of these elements, while prematurity will reduce the amount of time the fetus has to accumulate these stores, and if not compensated for will predispose them to deficiency diseases in the first few weeks of life. After birth, the newborn relies on the special qualities of breastmilk to provide them with the nutrients and immunological protection they need to thrive. Human breastmilk contains over 100 different constituents that are vital for the health, growth and development of the suckling child. Colostrum, the most nutritious part of the breastmilk, has a lemony-yellow viscous appearance and is present in the first few days after birth. It contains a high number of antibodies, immunoglobins (predominantly IgA) and the iron-bonding protein, lactoferrin (Hansen and Winberg, 1972; Jelliffe and Jelliffe, 1978; Fildes, 1986a), and although it contains less fat and lactose than mature milk, it has higher levels of sodium chloride and zinc (Jelliffe and Jelliffe, 1978). Colostrum is usually replaced by the thinner mature milk by the 10th day of breastfeeding (Fomon, 1967), but human milk will provide an immunological boost to the child and provide an uncontaminated source of protein throughout lactation (Dettwyler, 1995).

Human breastmilk is high in carbohydrates in the form of lactose, which enhances calcium absorption and prevents the child from developing diseases such as rickets (Jelliffe and Jelliffe, 1978). Trace elements in the diet are essential for the production of enzymes and key metabolic functions, and are readily available in human milk, allowing the child to gain sufficient amounts despite a reduction in the quantity of breastmilk with time (Chierici and Vigi, 1991). Two of the most important trace elements found in human milk are iron and zinc which are essential for growth. Iron is necessary for the production of haemoglobin and myoglobin, and fetal stores accumulated in the womb can be recycled in the infant for the first few months of life, before additional iron needs to be added to the diet. Anaemia in the mother may deplete the stores in the growing child, resulting in growth retardation and prematurity. In addition, consumption of food high in phytic acid, such as cereals, legumes, cows' milk and egg yolk, will inhibit iron absorption if not compensated for by meat or other haem foods (Working Group, 1994). Millions of years of natural selection have ensured that the constituents of human milk change as the child gets older, providing the appropriate mix of protein, energy, vitamins and minerals at each age (Dettwyler and Fishman, 1992). Quant (1996) argues that the low fat and high carbohydrate content of human breastmilk reflects the evolution of feeding in early humans. It dictates frequent feeding and intimate and regular contact between the mother and child, as well as causing hormonal changes in the mother that suppress ovulation and delay further pregnancies for around 9 months.

The IgA molecules passed to the suckling child are produced by the mother's immune response to pathogens in her surrounding environment, and hence, are tailor-made for the pathogens the child will encounter during the first few weeks of life (Newman, 1995). Passive immunity from diseases such as measles, rubella, polio, mumps, hepatitis B virus and tetanus is gained from a wider variety of immunoglobins transferred from the mother to the child in the womb. These transplacental antibodies provide the greatest protection at birth, but will be completely eliminated from the child's system by around 1 year (Hoshower, 1994). As they remain within the digestive tract, antibodies secreted in colostrum provide protection against enteric diseases including *Escherichia coli*, salmonella, dysentery and cholera (Hoshower, 1994). In addition, it is believed that some molecules within the breastmilk inhibit the supply of minerals and vitamins that harmful bacteria need to multiply in the gut, while others stimulate the infant's own immune system (Newman, 1995).

6.3   Weaning and infection

The Anglo-Saxon word for weaning is *wenian*, meaning 'to become accustomed to something different' (Katzenberg *et al.*, 1996). This seems to suggest a gradual process. 'Weaning' is a term that has been used to define both the introduction of additional foods to the infant diet, and the total and abrupt cessation of breastmilk (complete weaning). After 4–6 months exclusive breastfeeding no longer provides the child with the vital nutrients it needs to grow and develop, but weaning a child before 4 months can be harmful. It exposes the unprepared infant to new microbes, and a diet that may contain nutrients the child is not yet able to metabolise, thus increasing the risk of infant mortality (Foote and Marriott, 2003). For example, some studies have raised concerns about the introduction of foreign proteins in foods such as cows' milk that are able to penetrate the highly porous infant gut and stimulate the immune system. Their highly soluble nature overloads the child's kidneys with concentrated urine (Foote and Marriott, 2003). There is also evidence that cows' milk stimulates bleeding of the infant gut resulting in anaemia (Working Group, 1994). Cows' milk is also suspected to inhibit the absorption of zinc, which is vital for the child's immunity and growth (Penny and Lanata, 1995; Sazawal *et al.*, 1995). The modern debate over the use of cows' milk as a substitute for human milk has repercussions for our understanding of infant feeding in the past. Although cows' milk contains iron and other trace elements found in human milk, these are less bioavailable to the infant hence, they must be introduced in the complementary diet.

After 6 months, the transition from breastmilk to solid foods, although essential, exposes the child to an increasing number of bacterial and parasitic infections causing potentially fatal diarrhoeal disease and malnutrition. This situation is known as the 'weanling's dilemma' (King and Ulijaszek, 2000). Malnutrition has an intimate interaction with disease, affecting the integrity of the skin, which acts as a barrier to pathogens, causing hypoplasia of the lymphoid system and a reduction in cell-mediated immunity (Ulijaszek, 1990). This synergistic relationship is best illustrated by diarrhoeal disease, which among other things, inhibits the digestion of nutrients and causes anorexia. Malnourished children are at greater risk of infection and, although they grow at the same rate as their better-nourished counterparts for the first 6 months, they will eventually experience growth retardation (Scrimshaw *et al.*, 1959; Beisel, 1975; Chen, 1983) and the formation of Harris lines (Gordon *et al.*, 1963). As nutritional stress causes children to be less resilient to future infections, cycles of infection and malnutrition become 'causally intertwined' (Kuzawa, 1998:195). If left untreated, infants can suffer persistent diarrhoea and may eventually die of dehydration, cardiac arrest, septicaemia, pneumonia and shock (Kilgore *et al.*, 1995).

Today, studies in industrialised countries have shown that first-generation migrants often tend to abandon their traditional breastfeeding practices in favour of the more convenient and fashionable formula foods (Núñez-de la Mora *et al.*, 2005). Hildes and Schaefer (1973) examined the health of Inuit who moved from scattered hunting camps to 'urbanised' centres in Igloolik. They found that the adoption of modern formulas resulted in an increase in the incidence of middle-ear disease (otitis media) in the children. The abandonment of breastfeeding was a pattern also seen in Pakistani populations in Bradford, UK (Aykroyd and Hossain, 1967). Although adults maintained their own dietary habits, breastfeeding of infants was abandoned within the first few weeks, as opposed to the usual 1 year in Pakistan. Amirhakimi (1974) found that in Iran, children from high-income families were less commonly breastfed and were weaned earlier, at 5–6 months, compared to the urban poor and village children who were universally breastfed and weaned between 18 and 24 months. This pattern of infant feeding seems to reflect similar practices in the past.

6.4   Ancient feeding practices

Evolutionary biologists have attempted to examine the development and processes of pregnancy, lactation, infant nutrition and weaning in early humans

(Dettwyler, 1995; Dufour and Sauther, 2002; Robson, 2004). Kennedy (2005) suggests that early weaning in humans (at about 2.5 years) in relation to great apes (5–8 years) occurred due to the protein needs of the developing brain that could not be sustained by mother's milk alone after 2–3 years. While some authors have argued that the transition to agriculture provided populations with the means to wean their children early with increased access to cereal grains and animal milk (Sellen, 2001; Bocquet-Appel, 2002), others contest that foragers did not lack weaning foods, and that the impetus for early weaning cannot be explained by agriculture alone (Kennedy, 2005). The examination of dental microwear from a 4–5-year-old child from Mesolithic Abu Hureya in Syria led Molleson and Jones (1991) to suggest that at this relatively late age the child was consuming pre-chewed food of a softer consistency than that of the adults in the sample. Smith (1992) noted that the time of weaning in 21 species of primates coincided with the eruption of the first permanent molar, which would make the natural weaning time for humans 5.5–6.0 years. This is due, Smith (1992) suggests, to the need of the child to have the correct dentition to process adult foods. However, the appearance of a full set of deciduous teeth by around 3 years in human children negates this argument. In fact, the more traditional age of weaning at around 2.5 years corresponds with the complete eruption of the deciduous dentition. Dettwyler (1995) suggests that the eruption of the first permanent tooth may actually coincide with the maturity of the immune system at 6 years, when the child no longer needs the immunological boost provided by the breastmilk.

For the historic periods, Valerie Fildes amassed extensive literature on infant feeding practices, including the employment of wet-nurses, weaning foods and taboos, up to the introduction of commercial feeding formulas in the 1800s (Fildes, 1986a, b, 1988a, b, 1992, 1995, 1998). For example, from the eleventh century onwards in England, wet-nurses became fashionable among the wealthy, with children being nursed in the home or kept nearby (Fildes, 1988a). By the seventeenth century, wet-nursing was at its height in England and urban babies were sent as far as 65 km from the town to be nursed. Infants who were sent to the countryside had high mortality rates, with the rural parish of Ware in Hertfordshire reported to have buried 1400 children from London in this period (Fildes, 1988a). Children sent to the countryside would not only have been denied colostrum, but any transplacental protection (passive immunity) provided by the mother against the urban environment would have been useless in protecting against a new strain of rural pathogens. Conversely, by the time the child was weaned and returned to the city, the child's antigens would be adapted for the environment in which it had been weaned, and not the urban one to which it was returning. Fildes (1986a) suggested that if the ability of a mother to produce breastmilk was compromised in the past, the child would die unless a wet-nurse could be afforded. In medieval Europe, colostrum was considered harmful due to its different colour and consistency, and was often denied the newborn child. Instead, butter, oil of sweet almonds, sugar, honey, syrup or wines were fed to newborns to make them vomit (purge) and clear the mucus from their mouths and intestines. Milk fever was a serious health risk for mothers who did not clear their breasts of milk, and infection from spoons and dishes caused potentially fatal gastrointestinal diseases, which were only resolved when the child was finally put to the breast, 2–4 days after birth (Fildes, 1986a). During the medieval period, the popular weaning foods were 'pap' (flour and bread cooked in water) and 'panada' (flour or cereal in a broth with butter or milk). Tough, durable feeding vessels were favoured by mothers, and bottles were often not cleaned before the next feed or between babies, once again providing the perfect environment for infection (Thompson, 1984). Rare examples of feeding vessels have been found on sites and within infant graves throughout Europe, for example at Saxon Castledyke in Humberside, UK and medieval Hjemsted in Denmark, and from sites in Roman France (Coulon, 1994; Crawford, 1999). These suggest that not all infants were exclusively breastfed during their first year of life. In addition, artistic representations of cow horns being used to feed infants have been reported in medieval France, Holland, and Winchester Cathedral, UK (Hooper, 1996), but it is not possible to tell if these vessels represent supplementary or artificial feeding. The absorption of zinc supplied in the breastmilk was probably hindered by the high cereal content of the pap which may have led to growth retardation (Binns, 1998). Not surprisingly, medical writers associated weaning with disease such as rickets, gastrointestinal disease and growth retardation (Fildes, 1986a). In 1748, William Cadogan (1748, cited in Stone, 1977) listed the major causes of infant death as fever during teething, intestinal worms and an inadequate milk supply from the mother or nurse. Poisoning from pewter dishes and lead nipple shields was common, and mercury and tin purges were administered as a cure for worms.

In England in the seventeenth and eighteenth centuries, mean weaning age dropped from 18 months to 7.25 months, a reduction that is associated with migration of females to the cities to find work (Fildes, 1986a). Thompson (1984) suggests that the popularity of artificial feeding may have been due to poor mothers in the urban environment being unable to feed their babies due to severe malnutrition. However, this was probably rare as mothers will continue to produce breastmilk of adequate quality and quantity despite deficiencies in their own diet, unless they experience extreme nutritional stress (Dettwyler and Fishman, 1992). The development of a rural industry in the eighteenth century provided employment opportunities for women in textile factories with consequences on healthcare, breastfeeding and infant mortality.

The age of weaning was brought forward or breastfeeding was interrupted, due to unpredictable income and parental illness, with a subsequent increase in family size and reduced birth intervals (King, 1997). By 1895, condensed milk was widely available in England, when consumption of artificial food doubled. During the hot summer months, the risk of contamination once the tin was opened increased the incidence of infant diarrhoea. Cows' milk produced in rural and urban dairies was associated with the spread of tuberculosis, scarlet fever and cholera in both areas (Wilson, 1986; Atkins, 1992). In the nineteenth century, a change in attitudes towards breastfeeding meant that wealthy women once again fed their children, and the beneficial aspects of colostrum were recognised. Fildes (1986a) argues that this phenomenon contributed to a rapid decline in neonatal mortality.

6.5   The osteological evidence

In order for the signs of stress and malnutrition to be evident on the skeleton or dentition, an individual must first recover from the episode and resume normal growth. In cases of acute disease or prolonged stress, an individual will die before the body can recover. Hence, the presence of skeletal lesions 'implies a good immune response and a relatively healthy individual' (Ortner, 1991:10). For palaeopathologists, the absence of lesions associated with malnutrition or disease is often taken as an indication that the individual was healthy. A controversial paper by Wood and colleagues (1992) entitled 'The osteological paradox' highlighted this incongruity by questioning these basic and widely accepted assumptions about past health patterns. They argued that skeletal samples could only provide information on the morbidity and mortality of non-survivors within a population, but could never truly reflect the living healthy population from which they were derived. These comments provoked researchers to examine the methods and interpretative techniques typically employed in the discipline. Goodman's (1993) critical response highlighted the need to examine both the age-at-death and multiple indicators of stress in order to identify the disadvantaged within a skeletal sample.

Traditionally, high frequencies of stress indicators in children between the ages of 2 and 4 years have been related to weaning stress, and mortality profiles and growth curves are often constructed to identify this stressful period in a child's development. For example, Wall (1991) examined Amerindian children from pre-European contact Central California (4250–300 BP) and attributed a lag in their growth profiles between 2 and 3 years as indicative of weaning stress. The subsequent increase in diaphyseal lengths in the 3–4-year age category was argued to represent recovery and catch-up growth. Similarly, Molleson

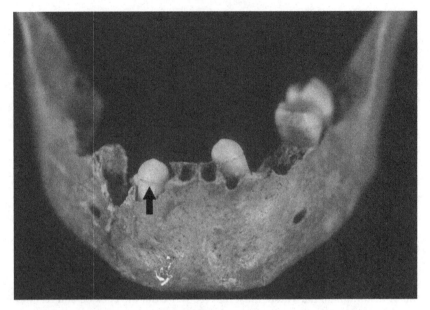

Figure 6.1   Enamel hypoplasias on the developing canines of a non-adult from
St Helen-on-the-Walls, York, UK (AD 1100–1550).

(1992) explained the small diaphyseal lengths of infants from Poundbury and
Spitalfields as representing early weaning onto cows' milk and pap, resulting
in zinc deficiency and growth retardation. Molleson (1992) goes on to argue
that the children never fully recovered from their growth retardation and as a
result became stunted in adulthood.

The aetiology of any single indicator of stress is complex, with many
different conditions known to contribute to its appearance (Goodman *et al.*,
1988). Therefore, it is essential to understand the aetiology and manifestation
of these lesions before assumptions about health and nutrition in the past can be
made.

6.5.1   Dental enamel hypoplasias

Dental enamel hypoplasias are areas of decreased enamel thickness that occur
during a disturbance of ameloblast deposition on the developing crowns of permanent and deciduous teeth (Fig. 6.1). It has been proposed that elevated corti-
sone levels, as a result of a stress episode, inhibit protein synthesis and reduce the
secretion of the enamel matrix (Rose *et al.*, 1985). The subsequent defects are

seen macroscopically as either furrows or pits, and are evident microscopically as irregular or exaggerated striae of Retzius (incremental growth lines), known as Wilson bands (Goodman and Rose, 1990).

Dental enamel hypoplasias provide a chronological and almost permanent record of stress episodes during the pre- and postnatal period (<7 years). Sarnat and Schour (1941) concluded that in 51% of the cases of hypoplasias in their modern sample, the exact cause of the defect could not be ascertained. Clinical studies have linked the defects to fever, birth trauma (Kronfield and Schour, 1939), congenital syphilis (Hillson *et al.*, 1998), low birth weight (Seow, 1992), severe childhood malnutrition (Sweeney *et al.*, 1971), rickets and hypocalcaemia (Kreshover, 1960; Levine and Keen, 1974), tuberculosis (Knick, 1982), zinc deficiency (Dolphin and Goodman, 2002) and intrauterine undernutrition due to a deficient maternal diet (Noren *et al.*, 1978; Acosta *et al.*, 2003). Hypoplasias on deciduous teeth often take the form of pits, or localised hypoplasias, particularly on the canines. First described by Skinner (1986; Skinner and Goodman, 1992), these defects occur in nutritionally disadvantaged groups and in children born during months of low sunlight and with low levels of retinol (i.e. vitamin A deficiency), which may hinder the growth of the alveolar bone. Thinning of the alveolar bone overlying the deciduous tooth may increase its susceptibility to trauma during birth, causing localised death of ameloblasts (Skinner and Newell, 2003). For example, these defects are common in the Sudan where teething problems are treated by lancing the alveolar process over the unerupted tooth with a heated needle (Lukacs and Walimbe, 1998; Lukacs *et al.*, 2001).

The age of the individual at the time a defect was formed is usually calculated by taking a measurement to the defect from the cemento-enamel junction (CEJ). These measurements are then converted into a chronological age by estimating the number of years it takes for the crown to be completed and into which half year of development the defect falls. A population specific method for estimating the time of defect formation involves taking the mean crown height (for each tooth being assessed) from the unworn teeth of individuals in the study sample, and applying the measurements to the following equation (Goodman and Rose, 1991:289):

$$
\begin{aligned}
\text{age of formation} = \ &\text{age at crown completion} \\
&- (\text{years of formation/crown height}) \\
&\times (\text{defect height from CEJ})
\end{aligned}
$$

Mandibular canines and maxillary central incisors are considered to be the teeth most susceptible to hypoplasias (Goodman and Armelagos, 1985; Santos and Coimbra, 1999). As canines take up to 4 years to develop in the jaw, they provide the widest time-frame in which to record a stress episode. The standard used to estimate the period of crown formation (e.g. Moorrees *et al.* 1963a, b) should be the same as that used to age the non-adult individuals if there is to be consistency between estimating the time of defect formation, and the age-at-death. In addition, it is important to be sure that the teeth are unworn, as attrition can obliterate the earliest lines. The use of posterior teeth and cementum hypoplasias on tooth roots has been advocated as a way to estimate stress during the later childhood years (Santos and Coimbra, 1999; Palubeckaité *et al.*, 2002; Teegan, 2004).

The age-at-death of individuals with stress indicators is often used to identify those most at risk from mortality in past populations. Most of these studies concentrate on hypoplasias in both the deciduous and permanent dentition. White (1978) noted that African hominids with dental defects were dying between the ages of 4 and 13 years as opposed to between 8 and 31 years in non-hypoplastic individuals. Since then, a number of researchers have found a significant correlation between enamel hypoplasia formation and a lower expectancy of life in adults (Rose *et al.*, 1978; Cook and Buikstra, 1979; Rudney, 1983; Blakey and Armelagos, 1985; Goodman and Armelagos, 1988; Duray, 1996; Stodder, 1997; Šlaus, 2000). The association of enamel hypoplasias to a lower mean age-at-death in adults has been taken as evidence that these features represent stress insults in susceptible individuals, and that they reflect cultural exposure that continued from childhood and adolescence into later life (Goodman and Armelagos, 1988; Palubeckaité *et al.*, 2002). Cook and Buikstra (1979) examined the deciduous dentition of children from the Lower Illinois Valley, USA. Their results showed children with defects that had developed in the prenatal period died earlier than those who developed hypoplasias after birth, suggesting selective mortality for individuals affected in the womb. This pattern was also seen in the Dickson Mound non-adults by Blakey and Armelagos (1985).

Numerous studies throughout North America and Europe indicate that the most frequent peak in the formation of enamel hypoplasias is between the ages of 2 and 4 years; this time period fits neatly with the age of weaning in most developing societies. Therefore, stress caused by a change in the quality and quantity of food during weaning has been implicated in the formation of these lines (Cook and Buikstra, 1979; Corruccini *et al.*, 1985; Goodman *et al.*, 1987; Lanphear, 1990; Wright, 1990, 1997; Moggi-Cecchi *et al.*, 1994; Lovell and Whyte, 1999). Doubts about this interpretation have been raised as studies began to show discrepancies of up to 4.5 years in the documented age of weaning, and that derived from examination of enamel hypoplasias formation (Blakey *et al.*, 1994; Wood, 1996). It is now considered that, during the 2–4-year developmental period, the enamel is more susceptible to environmental disturbance, and that the position of the defect on the tooth crown may be related to the structure of the enamel layers, with earlier defects being hidden by later appositional layers in the occlusal zone of the tooth. The morphology of the tooth means that, when recorded macroscopically, only lines formed at 2–4-years are visible (Hodges and Wilkinson, 1990; Hillson and Bond, 1997; Reid and Dean, 2000). Ameloblast exhaustion has also been cited as attributing to the appearance of these defects during this developmental period (Suckling and Thurley, 1984; Goodman and Rose, 1990). In addition, the width of the defects are determined by the positioning and spacing of the perikymata (external expression of Wilson bands) as well as the number of perikymata affected, rather than being due to the duration of the stress (Guatelli-Steinberg, 2003a, b).

These studies have discredited much of the earlier work that correlated peaks in enamel hypoplasias with weaning stress, so much so that Larsen (1997:49) considered the link between enamel defects and weaning to be nothing but 'coincidental'. New methods to combat these problems have been introduced. Dental defects are now more accurately and fully recorded using histological analysis which allows Wilson bands to be counted using 9-day intervals from the neonatal line (King *et al.*, 2002; Cunha *et al.*, 2004; Nava *et al.*, 2005). Fitzgerald and Saunders (2005) suggested that such counts are more accurate on deciduous teeth where there is less postnatal enamel and, as they all develop prenatally, they will all exhibit neonatal lines making the counts easier to calculate.

6.5.2   Harris lines

Transverse lines of increased radio-opacity, or Harris lines, occur in the growing bones of non-adults, and have remained a popular but problematic indicator of physiological stress in ancient populations (Fig. 6.2). These features were first discussed by Ludloff (1903), who observed them in the long bones of apparently healthy individuals, but it was not until 1931 that their aetiology was fully explored (Harris, 1931). Park described the precise mechanism behind the formation of the lines in the 1950s, after carrying out experiments on rats (Park and Richter, 1953; Park, 1954, 1964); it can be summarised thus:

(1) An episode of acute or chronic stress results in a deceleration in the development of the growth plate as the chondroblasts cease to lay down cartilage.
(2) The osteoblasts react more slowly to the stress and, for a time, continue to deposit osteoid along the static growth plate, producing a thin layer of bone along the transverse cartilage cap. This line is not visible on a radiograph.
(3) When normal growth resumes, the osteoblasts recover more quickly than chondroblasts and, once again, deposit osteoid on the cartilage cap, producing a thicker lattice of dense bone, substantial enough to be visualised on a radiograph.

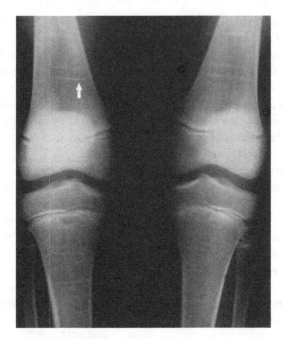

Figure 6.2 Multiple Harris lines in a child aged 11 years who had apparently been healthy. From Caffey (1978:599), reproduced with kind permission from Elsevier.

(4) This process continues until the maturation cycle of the chondroblasts has been completed. Hence, a line is only evident if normal growth is resumed and the individual recovers from the stress episode. Once the epiphyses have fused, these lines can no longer develop.

The exact aetiology of Harris lines has never been ascertained, but starvation (Park and Richter, 1953), septicaemia, pneumonia (Acheson and Macintyre, 1958), lead poisoning, rickets, congenital syphilis and scurvy (Follis and Park, 1952) are among the many conditions associated with a slowing of growth and development of the lines. Sontag and Comstock (1938) added emotional stress to this list, and Harris (1933) identified neonatal and fetal lines, which he associated with birth trauma and poor maternal health. Today, the lines are often associated with weaning (Dreizen *et al.*, 1964), and young children are thought to be more predisposed to the development of Harris lines due to rapid growth and frequent illnesses. Although nutritional status has only an indirect effect on the formation of Harris lines, good nutrition is directly related to the child's ability to recover rapidly from the stress episode and to begin catch-up growth. This process is implicated in the greater number of fragmentary (resorbed) lines in better-nourished children (Dreizen *et al.*, 1959).

Harris lines have been used to establish the frequency of stress episodes in a population by counting the mean number of lines, the percentage of individuals affected, and/or the age at which these lines occur most frequently. Several authors have developed methods to assess the age of the individual at the time of formation of Harris lines. One of the first attempts was by Allison and colleagues (1974), who measured the maximum length of the tibia, subtracted an arbitrary birth length (90 mm), and divided the remaining proximal and distal ends of the bones (i.e. the estimated growth area), into 16 equal parts representing 16 years of growth. This method was rather simplistic and assumed an equal annual growth rate and minimal male/female variability. In addition, the authors did not cite the source of long-bone length at birth, or their standard for epiphyseal fusion ages. Since then, various methods to age the formation of Harris lines have been published for the tibia (Clarke, 1982; Maat, 1984), the tibia and femur (Hunt and Hatch, 1981) and for the femur, tibia, humerus and radius (Byers, 1991).

Work incorporating Harris lines as indicators of stress has been hindered in adult skeletal populations due to the unpredictable and frequent remodelling of the lines. This was not always recognised. In 1964 Wells stated that 'except in certain relatively rare circumstances [Harris lines], are permanent' (Wells 1964:155). Now we know that the use of Harris lines to estimate the frequency of stress episodes is fraught with problems. Marshall (1968) argued that the number of lines on a child's radius did not relate to the number of insults they had suffered in the past. A year later, Gindhart (1969) showed that diseases were only followed by a line in 25% of cases, and 10% of the lines occurred when no stressful episode was documented. Males and females have different rates of remodelling, and stresses as mild as an injection or as severe as malnutrition may cause cessation of growth and the development of a line (Dreizen *et al.*, 1964). It has yet to be decided whether the thickness of a line is a measure of stress severity or longevity (Huss-Ashmore *et al.*, 1982), or whether it may be related to individual variability in the thickness of the cartilage plate or speed of chondroblast recovery. In fact, Park (1954) asserted that the position of the bone during radiograph might give a false impression of a thicker line. It should be remembered that the individual would need to recover from the stress in order to display a line. Therefore, weaker individuals, who could not rally from stress and regain normal growth, would not develop lines visible on a radiograph.

Factors associated with the development of Harris lines are so similar to those implicated in enamel hypoplasias that many attempts have been made to correlate the ages of formation of these defects (McHenry and Schulz, 1976; Clarke, 1982; Maat, 1984; Mays, 1985; Alfonso *et al.*, 2003). These studies have been mostly unsuccessful. Firstly, there may be a difference in the severity of the stress needed to disrupt the more environmentally robust enamel, and secondly, enamel does not remodel. In an adult sample, the earliest Harris lines will have been resorbed due to the normal remodelling process. Therefore, the frequency should be expected to diminish with age, rather than their absence in mature adults indicating individuals with a more adept immune system, as asserted by Nowak and Piontek (2002). Blanco and colleagues (1974) found that children in Guatemala suffering from chronic malnutrition displayed Harris lines and were shorter than those without lines, and reasonably suggested that these factors were linked. Studies carried out to test such a hypothesis have consistently shown that transverse lines are not associated with short stature (Mays, 1985; Carli-Thiele, 1996; Ribot and Roberts, 1996). These studies indicate that children were capable of a full recovery from the stress that caused the line and that catch-up is sufficient to prevent any growth retardation (see Chapter 4). Conversely, some authors have argued that development of Harris lines is actually associated with accelerated rates of linear growth, and is part of the normal growth process (Magennis, 1990). However, these studies have been unable to establish whether the periods of rapid growth velocity were due to catch-up growth after an insult, rather than being the causative factor. In recent times, Lampl and colleagues (Lampl *et al.*, 1992; Lampl and Jeanty, 2003) have shown that periods of growth stasis do occur as part of the normal growth process. This may explain the appearance of Harris lines when no period of illness has been recorded.

Problems with the aetiology of transverse lines aside, there are major limitations associated with the recording of these features. Hughes and colleagues (1996) argued that the type and side of bone examined could affect the prevalence of the lines recorded, and that the distal tibia showed lines more frequently and clearly than any other bone. In addition, while pairs of bones should show the presence of Harris lines, bones from the left side had greater numbers of them. Hence, although the percentage of individuals estimated to have suffered a stress would not be affected, the frequency of stress would be inaccurate. In order to get a fuller picture of the stress experience for that individual or population, radiographs of all the long bones would need to be taken. Due to the poor preservation of most skeletal remains, this would prove impossible. As most long bones develop through a rotating pattern of lateral and medial bone deposition (Garn and Baby, 1969), remodelled lines will be evident as fragments in the centre of the shaft. The tradition of scoring lines that extend 'halfway' across the bone, rather than counting all visible features, will lead to under-scoring. A confounding problem was demonstrated by a study carried out by Macchiarelli and colleagues (1994) who showed major discrepancies in observer error when counting the number of lines present on a bone. Further studies found a 50% disagreement in counts for their intra-observer study and 43% for the inter-observer counts (Grolleau-Raoux *et al.*, 1997). Hence, it may

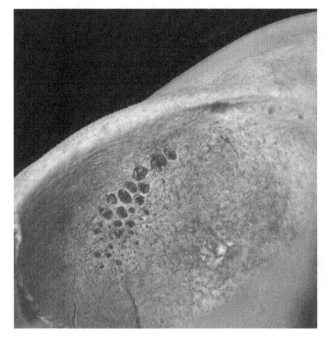

Figure 6.3 Cribra orbitalia in the right orbit. (Photograph: Jean Brown.)

be better to simply record the presence and absence of these lines in skeletal populations.

6.5.3   *Cribra orbitalia and porotic hyperostosis*

The term 'cribra orbitalia' refers to porous lesions on the orbital roof (Fig. 6.3) which are thought to be associated with similar lesions on the cranial vault known as 'porotic hyperostosis' (Fig. 6.4). These lesions result from an over-activity of the marrow, a thinning of the outer table of the skull, widening of the inner diploë and a spiculated 'hair-on-end' appearance of the trabecular structure (Britton *et al.*, 1960; Moseley, 1974; Ponec and Resnick, 1984). During childhood, the diploë contains red bone marrow; an expansion of this space is thought to result from the need of the body to produce and store more red blood cells in cases of anaemia. These lesions only develop in childhood when the diploë is already filled to capacity with haemopoietic marrow, placing pressure on the outer table of the skull. In adults, an increase in red blood cell production does not involve the entire marrow space (Larsen, 1997). If the anaemia is

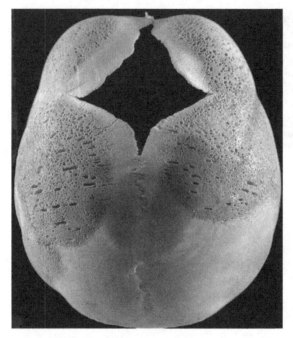

Figure 6.4   Porotic hyperostosis of the cranium. (Photograph: D. J. Ortner.)

resolved, the lesions will begin to remodel with the majority of studies showing healed lesions becoming more frequent with age. Therefore, non-adults will normally display the most frequent, active and severe lesions.

Cribra orbitalia describes a morphological feature, and although not a diagnostic term, through the years it has become synonymous with iron-deficiency anaemia. However, its aetiology was not always so clear-cut. Cribra orbitalia was first thought to be indicative of orbital haemangioma or chronic eye infections in leprous individuals (Møller-Christensen and Sandison, 1963; Duggan and Wells, 1964). Later studies suggested the cause was a nutritional deficiency (Angel, 1964; Nathan and Haas, 1966), most notably, iron-deficiency anaemia in the transition from a hunter–gatherer subsistence to a dependency on maize agriculture and the prohibitive affects of phytic acid (El-Najjar, 1977a, b; Saarinen, 1978; El-Najjar *et al.*, 1979; Dallman *et al.*, 1980; Cohen and Armelagos, 1984; Gilbert and Mielke, 1985). Chemical studies found low iron levels in affected individuals (Reinhard, 1992; Glen-Haduch *et al.*, 1997). Palkovich (1987) studied the prevalence of porotic lesions in non-adults from New Mexico and found remodelled lesions in children between the ages of 1 and 3 years. He argued that the maize diet resulted in malnourished mothers giving birth to

LBW babies, whose depleted iron stores resulted in anaemia by 6 months of age. Parasitic infestation, with subsequent intestinal bleeding, has also been implicated in the aetiology of cribra orbitalia and porotic hyperostosis (Reinhard, 1992; Trinkaus, 1977; Blom *et al.*, 2005), particularly in tropical regions where hookworm infections are common (Larsen, 1997). Walker (1986) argued that exposure to contaminated fish and water supplies resulted in weaning diarrhoea and iron deficiency due to malabsorption in a fishing community from California. Kent (1986) also suggested that the link between iron deficiency and its rise in agricultural communities was the result of sedentism and an increased pathogen load.

The premise for the 'pathogen load' theory is that as iron is a prerequisite for bacterial growth in the host, during an infection, the immune system withholds iron from invading micro-organisms as a defence mechanism (Weinberg, 1974, 1992; Brown and Holden, 2002). Interestingly, children with mild iron-deficiency anaemia have been shown to suffer fewer infections than those with a normal iron status or severe anaemia (Shell-Duncan and McDade, 2002). De la Rúa and colleagues (1995) returned to the question of dietary deficiency when they found a higher prevalence of cribra orbitalia in their agricultural populations than their industrial groups in a study of medieval populations in the Basque country. Fairgrieve and Molto argued that early weaning onto goats' milk led to cribra orbitalia at 6 months in the infants from the Dakhleh Oasis, and that anaemia was exacerbated by the use of honey as a weanling food, exposing the children to botulism, a theory in concordance with Holland and O'Brien (1997), who stressed the need for an approach that incorporates both diet and pathogen load when interpreting these lesions.

Today, the groups most at risk from developing iron-deficiency anaemia are non-adults between the ages of 6 months and 3 years, and women between the ages of 20 and 30 years. The highest frequency of the condition usually occurs before 5 years of age (Mensforth *et al.*, 1978; Goodman and Armelagos, 1988). Iron-deficiency anaemia is still one of the most common health problems worldwide, and it is estimated that 20–25% of the world's infants suffer from the condition. The combined effect of deficiencies in iron, vitamin A and zinc is linked to high child mortality, poor mental health and retarded motor skills, impaired growth and limited resistance to infection (Seymour, 1996; Ryan, 1997). Anaemia in infants is not the result of low maternal iron levels during pregnancy, as this has been shown to have little effect on the iron content in breastmilk (Dallman *et al.*, 1980). Instead, anaemia is thought to be produced by a combination of reduced iron levels in maternal milk after 6 months, and a depletion of the iron stores held by the infant at birth (Saarinen, 1978). Premature and LBW babies (under 2500 g) are susceptible to iron-deficiency anaemia before 6 months of age, as a result of lower iron stores and the greater need for iron during rapid catch-up growth (Schulman, 1959). Weaning onto cows' milk may exacerbate iron deficiency if it occurs before the child is 6 months old, due to intestinal bleeding from an irritation of the immature digestive system (Jelliffe and Blackman, 1962).

In palaeopathology, cribra orbitalia and porotic hyperostosis are usually graded according to the scheme described by Stuart-Macadam (1991:109) who provides photographs and a written description of the different appearances ('types'). An assessment of active (sharp-edged) and remodelled (smooth-edged, filled-in) lesions (Buikstra and Ubelaker, 1994) is useful when assessing the age at onset and relationship between these lesions and others, for example, non-specific infections. Jacobi and Danforth (2002) have raised the issue of inter-observer error when scoring the presence and state of healing of both orbital and cranial vault lesions. Some researchers do not make a distinction between the vault and orbital lesions and refer to both as 'porotic hyperostosis'. This term refers to the syndrome seen on radiographs of skulls by clinicians, which includes radiating trabeculae, thinning of the outer table, diploë and orbital roof thickening, orbital rim changes and the underdevelopment of the frontal sinuses (Sheldon, 1936; Jelliffe and Blackman, 1962; Stuart-Macadam, 1987). Orbital lesions are often seen in isolation, particularly in European samples. This has led some authors to argue that: (a) orbital lesions represent an earlier or milder form of iron-deficiency anaemia to that producing cranial lesions (Blom *et al.*, 2005); (b) the orbital and vault lesions found in North American populations are of a different aetiology to those, mainly orbital, lesions found in Europe; or (c) orbital and vault lesions are not of the same aetiology (Wiggins, 1996).

Just like cribra orbitalia, porotic hyperostosis is not pathognomic of iron-deficiency anaemia, and is also seen in the congenital forms of anaemia (e.g. thalassaemia and sickle-cell anaemia), and as a symptom of rickets and scurvy. Its aetiology can be distinguished from genetic anaemias by additional post-cranial lesions and DNA analysis (Moseley, 1974; Faerman, 1999; Ortner, 2003). In addition, cranial lesions associated with iron deficiency include frontal bossing, involvement of the ectocranial surface and the diploë (as opposed to the endocranial surface in rickets) and the absence of the lamina dura outlining the tooth socket (Lanzkowsky, 1968, 1977). As both rickets and scurvy are associated with anaemia, it is likely that all three conditions and their associated lesions will be present. Histological analysis is essential to differentiate between true cases of iron-deficiency anaemia, where gracile trabeculae are orientated in a perpendicular hair-on-end pattern extending from the diploic space, from porotic lesions caused by remodelled appositional new bone as the result of haemorrhage due to trauma or scurvy, tumours or inflammation (Schultz, 1997, 2001). In fact, in their histological analysis of cribra orbitalia in

333 individuals, Wapler and colleagues (2004) found no evidence of anaemia in 56.5% cases. Therefore, cribra orbitalia and porotic hyperostosis should be viewed as descriptive, rather than diagnostic terms (Ortner, 2003).

6.6 Weaning and bone chemistry analysis

6.6.1 Nitrogen isotopes

The study of weaning using bone chemistry analysis relies on the fact that nitrogen and oxygen stable isotope ratios, as well as elemental concentrations of strontium and calcium in bone, vary systematically according to the trophic level of an individual. Since the late 1980s, ratios of stable nitrogen isotopes with masses 14 and 15 ($^{14}N/^{15}N$ or $\delta^{15}N$) have been used to investigate weaning practices in the past. The body tissue $\delta^{15}N$ of a newborn infant is similar to that of its mother but as the child begins to breastfeed and effectively consumes its mother's tissues, it is raised to a higher trophic level on the food chain (Fogel *et al.*, 1989). Therefore during breastfeeding, and as the nitrogen from the breastmilk is incorporated into the offspring's body tissues, infant tissue $\delta^{15}N$ rises by around 2–3%. When weaning begins, and mother's milk is gradually replaced by other food, tissue $\delta^{15}N$ will decrease and eventually fall in line with the $\delta^{15}N$ values of the mother and other adults in that population (Katzenberg, 2000). Bone collagen is the tissue most frequently used to study weaning in archaeological populations. Because collagen turnover is relatively slow in relation to other body tissues there is a recognised, but as yet unmeasured, time lag between the nitrogen signature in infant tissues and their trophic status at a given time. For example, a child may be several months old before the trophic-level effect of breastfeeding which commenced right after birth is registered in the bone collagen (Larsen, 1997:285). This delay is also witnessed in modern infants under the age of 3 months, and it has been suggested that rapid growth of these infants affects the isotopic values to some degree (O'Connell and Hedges, 1999). When this theory was tested with regard to $\delta^{15}N$ values observed in the rapidly growing metaphyses, compared to the diaphyses in non-adult long bones, no difference in the nitrogen signature could be observed (Waters and Katzenberg, 2004).

In 1986, Buikstra and colleagues suggested that, with the introduction of agriculture and the development of ceramic vessels in North America, mothers had the ability to feed their children pap, resulting in earlier weaning ages, and hence, an increase in population as the contraceptive effects of prolonged lactation were withdrawn (Buikstra *et al.*, 1986). To address this question, Fogel and colleagues (1989) recorded $\delta^{15}N$ ratios in the fingernails of nursing mothers and their infants, and showed a decline in nitrogen values in infants as supplementary foods were introduced. When these analyses were carried out on pre-agricultural and agricultural sites, $\delta^{15}N$ decreased at 18–20 months. Later, Katzenberg and colleagues (1993) employed carbon and nitrogen isotopes to illustrate the use of maize in the weaning diet in the MacPherson Iroquoian village site in Ontario (AD 1530–1580), and used the evidence to support the theory that circular caries in children from the site resulted from subsequent weaning stress due to their high carbohydrate diet. In a later study, Schurr and Powell (2005) challenged the argument that the onset of agriculture resulted in new weaning foods and reduced lactation times. They examined the nitrogen and carbon isotope ratios of children from two pre-agricultural and two high agricultural sites in the prehistoric Lower Ohio valley and found no difference in weaning times between the two types of site, stating that weaning occurred in both between the ages of 2 and 5 years.

In 1997, Schurr had compared the age of weaning estimated from isotopic signatures with that indicated by infant mortality, which peaked between birth and 6 months at the Angel site. Schurr found no direct correlation, suggesting that the traditional indirect measures of weaning stress (i.e. age-at-death, indicators of stress) could not be relied upon (Schurr, 1998). Since this study bioarchaeologists and bone chemists have continued to refine data on the age of weaning in order to test its relationship to mortality. At Wharram Percy in later medieval Yorkshire, weaning is reported to have occurred at around 2 years (Richards *et al.*, 2002; Fuller *et al.*, 2003), with children up to the age of 11 years having a diet lower in animal protein than the adults. There was no evidence that the weaning diet differed between those that died as children and those that grew up into adulthood (Fuller *et al.*, 2003). Osteological evidence from the site indicates that enamel hypoplasia formation peaked at 1.5 years, but that child mortality, cribra orbitalia and non-specific infections peaked much later, indicating that weaning was successful in maintaining a healthy child (Lewis, 2002a). Katzenberg and Pfeiffer (1995) showed abrupt weaning at 1.5 years in children from the Methodist cemetery at Newmarket, Ontario – an age consistent with the documentary sources. In some cases there was no evidence that the children had been breastfed at all and this may be due to their death before the breastfeeding signal had registered in the bone. Other sites have shown a more gradual introduction of weaning foods, beginning at 3–4 years in ancient Nubia (White and Schwartz, 1994), 1–2 years at the Angel site (Schurr, 1997; Schurr and Powell, 2005), 9 months at MacPherson (Schurr, 1998) and as early as 6 months in Roman Egypt (Dupras *et al.*, 2001). At Kaminaljuyú in Guatemala, children were gradually introduced to maize at 2 years, but continued to be breastfed for up to 6 years (Wright, 1998), fitting well with Dettwyler's (1995) estimation of natural weaning age at between 2.5 and 7.0

years. Herring and colleagues (1998) examined collagen $\delta^{15}N$ in the non-adult sample from St Thomas' Church, Belleview coupled with skeletal age, census data and the parish records. They demonstrated a decrease in $\delta^{15}N$ ratios at 5 months as weaning foods were gradually introduced but that breastmilk continued to be a major source of protein for up to 14 months.

Our ability to measure an increase and decline of nitrogen in the infant has shown potential for indicating children who may have died at certain points in the weaning process (Williams *et al.*, 2002). It is also possible to measure isotopic signatures in archaeological hair to reconstruct diet (Macko *et al.*, 1999). Preserved hair is rare in archaeological burials, and there is a known lag of *c.*7–12 months for hair keratin to reflect a dietary isotopic shift (O'Connell and Hedges, 1999), meaning we may miss short period changes, especially if the hair was cut eliminating the sections that relate to weaning age. Thus far, despite studies in this area on adults, no research has been published on archaeological child hair samples, even though in the UK, archaeological hair is preserved at Christ Church Spitalfields, Poundbury Camp and in some royal child burials.

With such exciting developments it is easy to forget the many issues that may affect the validity of these studies. For example, we rarely have information on an individual's diet, state of health or cause of death, all of which are known to affect protein turnover rates (Katzenberg and Lovell, 1999). Katzenberg (2000) also warns that, as nitrogen levels are affected by malnutrition, and all the children in the archaeological record died, it may be impossible to differentiate between elevated nitrogen as the result of pathology and that due to the breast-feeding effect. Children may also have been given a different diet before their death if they suffered from an illness (Dupras *et al.*, 2001). Hence, researchers have begun to examine the permanent dentition, as these tissues are formed during infancy and turn over very little (dentine) or not at all (enamel), therefore retaining weaning signatures in older children and adults (i.e. the survivors of weaning).

6.6.2   Oxygen isotopes

Carbon isotopes are used to identify the introduction of cereal based weaning foods to the infants diet, with $\delta^{13}C$ increasing at the age when, and if, C_4-based foods are introduced. Oxygen, on the other hand, reflects the water source. Body water has a higher $\delta^{18}O$ than ingested water as ^{16}O relative to ^{18}O is lost during climatic water evaporation (Katzenberg, 2000). Non-breastfed infants will have lower levels of $\delta^{18}O$ than breastfed infants from birth and, as children are weaned and begin to ingest water from foods other than that contained in breastmilk, their $\delta^{18}O$ levels should decrease. Wright and Schwartz (1999) demonstrated that this pattern was observable in archaeological samples, and reported that with age $\delta^{13}C$ increased as $\delta^{18}O$ decreased. Because dental enamel is made up of incremental layers, the age at which weaning begins can potentially be precisely measured by targeting specific incremental lines. Macpherson (2004) correlated dental features including dental wear, caries and calculus with the age of weaning derived from oxygen isotopes, to suggest that infants from two sites in Anglo-Saxon England were weaned on to a low cariogenic diet before the age of 3 years. Oxygen values in dental enamel formed during and after weaning may fall below the adult average, due to maternal values affected by seasonal availability of food and water supplies, or maternal migration from a low to a high oxygen area (White *et al.*, 2004). Oxygen isotopes are usually measured in the enamel phosphate (Wright, 1998). The use of dental enamel and dentine, as opposed to bone collagen, has the advantage of allowing a more detailed pattern of weaning to be identified (Wright, 1998), although dentine signatures may provide a slightly older age signal than the enamel when sampled together (Wright and Schwartz, 1999). Laser ablation, which allows isotopic signatures to be taken from microscopic enamel or dentine incremental layers, is the most precise way of assessing the age of children making the transition from exclusive breastfeeding to full weaning (Fuller *et al.*, 2003; Humphrey *et al.*, 2004).

6.6.3   Strontium, calcium and zinc

Elemental concentrations of strontium and zinc have also been suggested to vary systematically during pregnancy and lactation. Strontium is discriminated against in relation to calcium in the placenta and in the synthesis of breastmilk, hence, breastfed infants will have lower strontium-to-calcium (Sr/Ca) ratios when compared to weaned individuals. A wheat and barley weaning diet will produce yet higher Sr/Ca ratios. Sillen and Smith (1984) were among the first to employ this theory to the question of weaning in a coastal Arab population at Dor, in the East Mediterranean. Their results suggested that most children were weaned by the age of 2–3 years. The use of cortical bone meant that precise measures of weaning age could not be estimated. Hühne-Osterloh and Grupe (1989) examined Sr/Ca ratios and zinc (Zn) concentrations in the bone mineral of children from medieval Germany, and concluded that weaning began at 6 months, when Sr/Ca levels rose and Zn levels decreased, reflecting the reduction of human milk, high in zinc. This weaning age was younger than the 2 years suggested by the medical literature at the time, and the authors suggested that the youngest infants were gradually receiving animal milk in place of human milk. They linked their chemical analysis to the occurrence of cribra orbitalia, Harris lines, rickets and scurvy in children as young as 18 months. The problem with using Sr/Ca values to reconstruct diet is that bone mineral is susceptible to diagenetic exchange with elements from the burial environment, which may alter the in vivo Sr/Ca signal. These processes are facilitated by the action of groundwater, and at present it is not possible to say with absolute certainty whether elemental concentrations in a bone have been diagenetically altered or not (Schurr, 1998). In fact, bone has been described as an excellent 'Sr sponge' due to its porous nature, fine crystals and the pockets and voids that facilitate the easy deposition and mobilisation of elements from the soil (Radosevich, 1993).

6.7 Specific diseases of malnutrition

6.7.1 *Rickets and osteomalacia (vitamin D deficiency)*

Rickets usually results from a deficiency in the hormone commonly known as vitamin D (1,25-dihydroxyvitamin D or $1,25(OH)_2D_3$). The majority of vitamin D is produced internally by the interaction of ultraviolet light with 7-dehydrocholesterol in the deep layers of the skin. Small amounts of vitamin D are present in foodstuffs such as fish oil and egg yolks (Fraser, 1995). Dietary deficiency in phosphorus and calcium has also been implicated in the development of the condition but the underlying cause is difficult to identify from skeletal lesions alone (Ortner, 2003). Nonetheless, vitamin D status in humans is primarily maintained by exposure to sunlight, which is required to promote the absorption and mobilisation of calcium and phosphorus from previously formed bone, as well as the maturation and mineralisation of the organic matrix (Pitt, 1988). Rickets refers to the effect of vitamin D deficiency on the chrondrocytes and growth plate, whereas osteomalacia reflects the effect of the condition on the osteoblasts, and their formation of osteoid and osteocalcin in bone modelling and remodelling. Hence, children can suffer from both conditions but, with the fusion of the epiphyses, individuals can only display osteomalacia (Pettifor, 2003).

At birth, the fetal skeleton comprises *c*.20–30 g calcium, making up to 98–99% of the total body mineral of the infant. Skeletal growth begins from mid-gestation and reaches its height during the third trimester. This rate slows at birth and again declines after the first few months after birth (Prentice, 2003). As vitamin D is passed to the fetus in utero, maternal vitamin D deficiency has been associated with congenital rickets, and may cause a delay in the appearance of the epiphyses (Prentice, 2003). It has also been shown that children can maintain high calcium levels despite hypocalcaemia in the mother (Kovacs, 2003). After 6 months, the stores of vitamin D begin to deplete and the mother and child need to ensure exposure to sunlight to make up for any dietary deficiency of the vitamin (Foote and Marriott, 2003). Premature babies miss out on the last 2 months of fetal life when the storage of calcium salts occurs (Arneil, 1973), making them more susceptible to hypocalcaemia, loss of bone mass and reduced growth in later childhood (Bishop and Fewtrell, 2003). In addition, LBW and twinned babies are also at risk due to their greater need of calcium and phosphorus as the result of catch-up growth after birth.

Chick (1976) reported that mothers of rachitic children had lower levels of the vitamin in their breastmilk than those without, indicating that the maternal body first looked after its own requirements before releasing the vitamin into the milk. The high calcium needs of the fetus, particularly at 27–28 weeks' and 37–38 weeks' gestation, must be compensated for by increased intestinal absorption and/or decreased renal excretion in the mother (Al-Senan *et al.*, 2001). Prolonged breastfeeding without vitamin D supplementation or adequate exposure to sunlight is a predisposing factor for the disease (Baroncelli *et al.*, 2000; Rowe, 2001; DeLucia and Carpenter, 2002) and today, the prevalence of rickets peaks at 3–18 months (Pettifor and Daniels, 1997). Early sources suggested that, in the past, a reoccurrence of rickets around the age of 6 years was not uncommon (Harris, 1933). Paradoxically, if a child is suffering from marasmus (severe protein–calorie deficiency) and has retarded growth, the signs and symptoms of rickets will not appear unless the marasmus is cured (Griffith, 1919). Hence, in the past, children suffering from a suite of malnutrition diseases may not show any visible signs of the disease on the skeleton.

Some cases of rickets are the result of defects in the individual's metabolism and are not indicative of dietary or environmental change. Longstanding renal failure, chronic acidosis, hepatic or pancreatic disease can all lead to rickets (Caffey, 1978), but were probably not compatible with life in the past. Vitamin-D-resistant rickets (refractory rickets) is a congenital condition requiring massive doses of vitamin D for treatment and can lead to dwarfism in adults (Holt, 1950). Abnormal bowing of the long bones has been reported in poorly reduced midshaft fractures, and bowing due to faulty fetal positioning can persist for up to 2 years after birth (Borden, 1974; Caffey, 1978). Today, the increasing prevalence of rickets in countries with high levels of sunlight, and 'Asian rickets' in British immigrants, has brought this disease to the forefront of research once more, where full body clothing, low calcium diets and darker skin limit the amount of ultraviolet light exposure (Narchi *et al.*, 2001). The modern incidence of rickets has raised new questions about the synergistic effects of lead, zinc and calcium metabolism in the aetiology of the condition (Thatcher *et al.*, 1999).

The clinical changes of rickets were first described in Germany by Whistler (1645), but it was not until the twentieth century that the work of Mellanby (1919) and McCollum *et al.* (1922) isolated the cause of the condition. Prior to the Industrial Revolution, rickets was a disease of affluence, where wealthy children were kept fully clothed and remained indoors (Pettifor and Daniels, 1997). With the rise of industrialisation, urban overcrowding and air pollution rickets became epidemic in northern Europe, and was so rampant in Britain that it became known as the 'English disease' (Mankin, 1974). In 1634, 14 deaths from rickets had been documented in London but, by 1659, as many as 1598 deaths were recorded (Fildes, 1986b), and in 1773, 20 000 children were reported to have been affected in the city (Harvey, 1968). In 1889, the British Medical Association carried out a survey on the distribution of rickets in Britain and found the disease to be prominent in the industrial and mining areas. As their survey moved further away from the urban centres, the frequency of the disease declined (Hardy, 1992). In 1890, Palm noticed that rickets was more common in cities where people were deprived of sunlight, and was the first to suggest that exposure to sunlight might be used to treat the condition (Palm, 1890). During the 1900s, Morse diagnosed rickets in 79.5% of 400 Boston infants and, in Germany, 89% of Dresden infants aged between 2 and 4 years had the disease (cited in Griffith, 1919).

In 1919 Mellanby, a British nutritionist, noted the effective use of cod liver oil in treating the signs and symptoms of rickets, and hence proved it to be a deficiency disease. This finding fitted with previous observations in urban areas, where Jewish children appeared to be unaffected by the disease, and their diet, made up predominantly of fried fish, may have provided them with the protection they needed (Hardy, 1992). The quality of breastmilk was another factor thought to contribute to the incidence of the disease. Parry (1872) claimed that the idle lifestyle of the rich, and the living conditions and excessive exercise in the poor, resulted in low-quality breastmilk. The lower prevalence of the disease in middle-class children was, Parry argued, due to good-quality breastmilk, a balanced diet and appropriate levels of exercise. In addition, poorer children could play outside where they were exposed to ultraviolet light, and they drank cows' milk, which was shunned by the rich as a food for invalids and infants (Fildes, 1986b). Rickets was closely associated with whooping cough and measles in the nineteenth century and this contributed to the high mortality rates from the disease (Hardy, 1992).

The manifestation of rickets and osteomalacia on the non-adult skeleton is related to the general nutrition of the child, the age at which the condition occurs, the child's rate of growth and its mobility (Harris, 1931). Uncalcified osteoid laid down on the growth plate and during the remodelling process, causes the bones to 'soften' and they become susceptible to bowing deformities.

Defective mineralisation of osteoid during the remodelling process is seen as Looser's zones (pseudofractures or Milkman's fractures), which are pathognomic of osteomalacia in clinical radiographs (Adams, 1997). However, one of the earliest signs of the disease is in the skull. The cranium becomes squared (craniotabes) on the side on which the infant lies, as a result of pressure on soft bone (Mankin, 1974). In addition, the deposition of osteoid on the external table of the skull mimics porotic hyperostosis, and the anterior fontanelle may remain open until the 3rd year. Enamel hypoplasias are common and dental development is delayed (Sheldon, 1943). This delay varies in its extent between individuals and between teeth, which is a matter of concern when trying to estimate the age of affected children within a skeletal sample (Liversidge and Molleson, 1995). For example, Holt (1909) carried out a survey on dental eruption in 150 rachitic children from a London hospital, and found that in 20 cases the dentition did not begin to erupt until the 12th month (as opposed to the 9th), and in eight cases dental eruption was delayed until 15 months.

In the undernourished rachitic child, the cortices are thinned and marrow spaces enlarged with general atrophy, and other conditions, such as scurvy, may be present. In a well-nourished child, rickets causes thickened cortices and a reduced medullary cavity (Stuart-Macadam, 1989). These children are described as having 'plump bones' as the long bones, ilia and scapulae thicken under the periosteum, as 'several layers of friable, spongy vascular tissue, with large vascular spaces' (Griffith, 1919:588). The long bones are susceptible to fracture, and there is a general retardation of growth and osteopenia, with excessive proliferation of cartilage at the metaphyseal ends of the bone that fails to be calcified. The wrist is usually the first area where changes to the growth plate are noticed on a radiograph. The metaphyseal ends become widened and concave as a result of weight bearing (trumpeting), the cortical bone is thinned and the metaphyses develop frayed edges, resembling 'bristles of a brush' (Caffey, 1978). Macroscopically, the characteristic smooth epiphyseal surface of the metaphysis is lost. In 2000, Thatcher and colleagues published a scoring method for rating the severity of these changes on a radiograph in the wrist and knee (Fig. 6.5) and such a scheme would be useful in the assessment and comparison of rickets severity in archaeological samples, although it has yet to be employed.

Children who maintain good muscle tone and remain mobile during the course of their disease have been shown to develop more severe bowing, cupping and flaring deformities than those with poor muscle tone who remain stationary (Stuart-Macadam, 1988). In the infant, bowing of the arms, where the humeral head is bent medially and inferiorly, suggests the child may have been crawling at the time of onset. Anterior bowing of the tibia may occur in response to the child lying with one leg resting upon the other (Pettifor, 2003).

WRIST[a]—score both radius and ulna separately

Grade	Radiographic features
1	Widened growth plate, irregularity of metaphyseal margin, but without concave cupping
2	Metaphyseal concavity with fraying of margins

2 bones × 2 points = 4 points possible

KNEE[a]—score both femur and tibia separately

Multiply the grade in A by the multiplier in B for each bone, then add femur and tibia scores together

A	Grade	Degree of lucency and widening of zone of provisional calcification
	1	Partial lucency, smooth margin of metaphysis visible
	2	Partial lucency, smooth margin of metaphysis not visible
	3	Complete lucency, epiphysis appears widely separated from distal metaphysis
B	Multiplier	Portion of growth plate affected
	0.5	≤ 1 condyle or plateau
	1	2 condyles or plateaus

2 bones × 1 point × 3 points = 6 points possible

Total: 10 points possible

[a]Score the worst knee and the worst wrist.

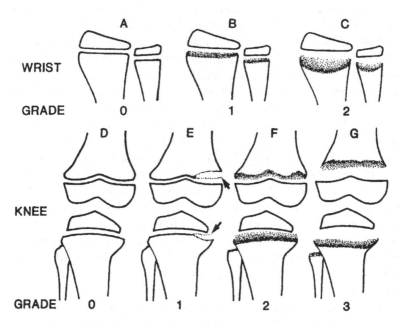

Figure 6.5 Radiographic scoring method for the assessment of rickets. A, normal wrist; B, irregularity and widening at the growth plate, but without cupping; C, metaphyseal cupping and frayed margins; D, a normal knee; E, only medial portions of the femoral and tibial metaphyses are affected; there is partial lucency of the metaphyses, but the margins are clearly visible (arrow); F, partial lucency of the metaphyses, but the margins are not sharply defined; however, zones of provisional calcification are not completely lucent and display some calcification; G, complete lucency of the zone of provisional calcification; the epiphyses appear widely separated for the distal metaphyses. From Thatcher *et al.* (2000:133), reproduced with kind permission from Philip Fischer and Oxford University Press.

An abnormal anterolateral concavity of the femoral shaft, a bowed tibia and anteroposterior flattening of the fibula may indicate that the child had begun to walk (Stuart-Macadam, 1988). Children with rickets have a habit of sitting cross-legged whist rocking to and fro, and this posture may contribute to the deformities (Holt, 1909). In all the affected bones, bowing is a result of weight-bearing and strain on a weakened growth plate (Caffey, 1978). Older children develop 'knock-knee' deformities (genu valgum) and severe cases may develop 'windswept' deformity, where one leg is bent towards the midline (varus) and the other away from it (valgus). Deformities of the femoral neck may also be present (coxa vara) (Pettifor, 2003). Hence, examination of bowing deformities can aid in assessing an age of onset for the condition. Subperiosteal new bone, stemming from abnormal osteoid deposition and accentuated by healing calluses from stress fractures in the osteopenic bone, may form on the concave surface, particularly the anterior aspect of the ribs, posterior surface of the femur and medial and posterior aspects of the tibia (Ortner, 2003).

The dual action of an unmineralised growth plate and bone softening can result in severe deformities of the rib cage. Cupping deformities present on the costal ends of the ribs from an abnormal accumulation of osteoid are known as the 'rachitic rosary' (Resnick and Niwayama, 1988). Pressure of the arms and the strain of the diaphragm in breathing results in a constricted 'pigeon chest', where the costochondral junctions are inwardly displaced (Caffey, 1978). Affected children have difficulty in breathing, and it is for this reason that they are believed to be particularly susceptible to respiratory infections such as pneumonia (Pettifor, 2003). In infants, rib fractures due to general osteopenia may be the first clinical sign of this disease, and should not be mistaken for abuse (Pettifor and Daniels, 1997). In severe cases, the vertebral bodies may collapse and develop kyphotic deformities. Schamall and colleagues (2003) took computerised tomography scans of the lumbar vertebrae of children from Vienna, known to have died while suffering from rickets, and noted thickened and irregularly arranged trabeculae, with enlarged cavities in the centre of the vertebral bodies causing partial collapse and reduced height of the bodies. Table 6.1 and Fig. 6.6 illustrate the type and distribution of lesions associated with rickets in the child.

While vitamin D deficiency is more easily identified on non-adult skeletons due to their rapid growth and the amount of uncalcified material lain down, this growth also means that signs of the disease can disappear just as quickly. Within 2–3 days of treatment, a zone of dense calcification appears at the end of the bone. The radiolucent rachitic metaphysis is remodelled and becomes normal in appearance as the calcification band is gradually incorporated into the diaphysis (Harris, 1931; Caffey, 1978). The gross enlargements at the joint ends become 'trimmed' and there is a period of intense remodelling, similar

Table 6.1 *Skeletal lesions in non-adults associated with rickets and osteomalacia*

Cranium	Postcranium	Radiographic features
Delay in fontanelle closure (Mankin, 1974)	Frayed metaphyses of long bones (esp. radius and ulna) (Caffey, 1978)	Radiolucent bands at metaphyses (Harris, 1933; Caffey, 1978)
	Thickened long bones, ilia and scapula (Stuart-Macadam, 1989)	Looser's zones (Adams, 1997)
Delay in primary and secondary dental eruption (Holt, 1909; Sheldon, 1943)	Rachitic rosary (Resnick and Niwayama, 1988)	Loss of dental lamina (Adams, 1997)
Parietal and frontal bossing (Ortner, 2003)	Anterior bowing of sternum (Pettifor, 2003)	
Craniotabes of occipital (Mankin, 1974)	Windswept deformity of femur and tibia (Pettifor, 2003)	
Dental enamel hypoplasias (Sheldon, 1943)	Anterior bowing of tibia (Pettifor, 2003)	
	Anteroposterior flattening of fibula (Stuart-Macadam, 1989)	
	Coxa vara of femoral neck (Pettifor, 2003)	
	New bone formation on anterior aspect of ribs (Ortner, 2003)	
	Reduced height of vertebral bodies (Schamall *et al.*, 2003)	

to catch-up growth (Fig. 6.7). All signs of the disease may have disappeared within 3–4 months (Harris, 1931). If the child is not treated and survives over the age of 4 years, stunting and deformities will persist (Sheldon, 1943).

Despite the endemic nature of rickets in the Industrial Revolution period, the diagnosis of this condition has, until recently, been neglected, with older studies recording the prevalence of enamel hypoplasias as evidence for the disease (Wells, 1975b). The early changes of rickets are subtle, and usually begin with an expansion and fraying of the rib ends and distal radius. The more well-known extreme bowing deformities are later manifestations, and occur when the child applies weight to the affected limbs. The disease can render children immobile and, as they are susceptible to whooping cough and gastrointestinal infections as a result of the disease, they may not live long enough to develop the most obvious signs. These factors may explain why so few cases of the condition have been reported in the archaeological record. Nevertheless, diagnostic techniques are becoming more sophisticated as the subtle changes to the ends of growing

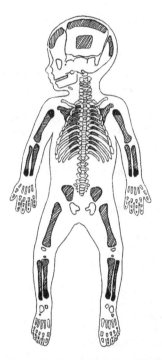

Figure 6.6   Distribution of nutritional rickets in the non-adult skeleton. Black indicates most commonly affected areas; hatched lines indicate areas of less frequent involvement.

bones and the ribs are studied in detail, and larger samples of medieval and post-medieval remains are being excavated or re-examined (Dawes and Magilton, 1980; Gruspier, 1989; Molleson and Cox, 1993; Lilley *et al.*, 1994; Ortner and Mays, 1998; Lewis, 2002c; Ortner, 2003).

6.7.2   *Infantile scurvy (vitamin C deficiency)*

Scurvy results from a dietary deficiency of ascorbic acid, or vitamin C, present in a wide variety of fruits and green vegetables. Unlike other mammals, with the exception of guinea pigs, humans and other primates cannot synthesise this vitamin internally and it must be ingested (Stuart-Macadam, 1989). Vitamin C is necessary for the formation of proline and lysine, amino acids vital for the synthesis of Type 1 collagen, which forms the basis of connective tissues for the skin, blood vessels, cartilage and bone. Vitamin C also protects and regulates the biological processes of other enzymes (Stuart-Macadam, 1989).

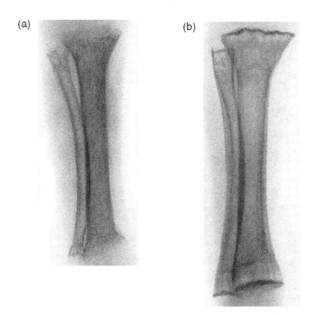

Figure 6.7 (a) Florid rickets in a on year old child. Note 'bristles-of-a-brush' appearance at the ends of the bones and general bone thinning (osteopenia). From Harris (1933:97), reproduced with kind permission from Oxford University Press. (b) Healed rickets in the same child after 14 days' treatment with cod-liver oil; note area of radiolucency above the growth plate. From Harris (1933:99), reproduced with kind permission from Oxford University Press.

The disease is clinically recognisable when the vitamin has been deficient for 4–10 months, or in children, once birth stores have been depleted (Ortner, 2003). Hence, scurvy is most common between 6 months and 2 years of age. As with rickets, premature, LBW babies and twins are susceptible (Griffith, 1919).

Before the sixteenth century, scurvy was common in soldiers and sailors on long sea voyages, but it was not recognised as a childhood disease until the rise of urbanisation. Initially described under the same umbrella as rickets, infantile scurvy or Möller–Barlow disease was not considered a separate condition for many years (Möller, 1862; Barlow, 1883), perhaps because the skeletal lesions of scurvy only appeared in advanced cases and were masked by the lesions of rickets (Barlow, 1935; Follis and Park, 1952). Wilson (1975) suggests that infantile scurvy was not recognised because the swollen, bleeding gums so characteristic of the adult disease did not occur in children before their teeth began to erupt. In the sixteenth century, as the potato became popular and market gardens more common, cities were supplied with fresh produce and scurvy declined (Lomax, 1986). In the eighteenth century, scurvy was common in the wealthiest families who could afford to feed their children on fashionable

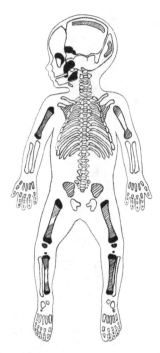

Figure 6.8 Distribution of scurvy in the non-adult skeleton. Black indicates most commonly affected areas; hatched lines indicate areas of less frequent involvement.

formulas including pasteurised and condensed milk, where processing destroys the small amounts of vitamin C contained within it (Harris, 1933; Lomax, 1986).

Scurvy is characterised by the primary effect of ascorbic acid deficiency on the collagen matrix, and the secondary effects of trauma to weakened bone and vessels. Minor trauma results in haemorrhage of the defective connective membranes and bleeding into neighbouring tissues. The distribution of lesions is summarised in Fig. 6.8. Fractures of the cortical bone are also evident at the metaphyses, as the bone matrix is undermined. In 1933, Harris carried out a survey of lesions in scorbutic children under 3 years of age and found that evidence of haemorrhage on the skeletal system occurred in 93% of cases and, in the orbits, 37% of cases showed either bilateral or unilateral lesions (Fig. 6.9). Scorbutic anaemia develops as abnormal grey marrow replaces red bone marrow at the end of the long bones causing arrested growth (Ortner and Putschar, 1985). Eventually, internal bleeding may lead to shock and, finally, heart failure.

In the 1980s, cases of scurvy were rarely reported in the archaeological record. Maat (1982) reported dark stains on the well-preserved dental roots and joint surfaces of Dutch whalers from eighteenth-century Spitsbergen Archipelago, which were later confirmed to contain remnants of haemoglobin

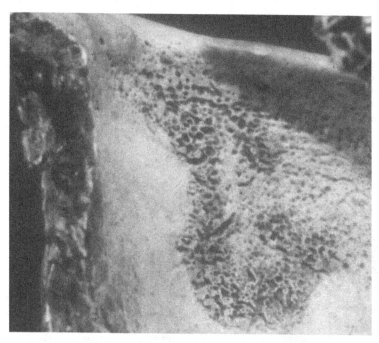

Figure 6.9 New bone formation in the orbit of a 3–4-year-old child from Beckford, Worcestershire, UK (100 BC–AD 100). (Photograph: Jean Brown.)

from scorbutic haemorrhage (Maat and Uytterschaut, 1984; Maat, 2004). Ortner (1984) described lesions of the skull in a child from Alaska, which he suspected as being indicative of scurvy, and this was followed by a case in a young Iron Age child from Beckford, Worcestershire, UK displaying new bone formation around a deciduous maxillary molar, the mandible, tibia and in the orbits (Roberts, 1987). Using histological techniques to distinguish haemorrhage from anaemia and inflammation, Schultz (1989) diagnosed scurvy in 13.8% of infants from Bronze Age Europe and Carli-Thiele (1995) recorded scurvy in 40% of a Neolithic population in Germany. Less convincing was the diagnosis of scurvy as the cause of a partial trephination in the skull a child from Middle Bronze Age Israel, who exhibited mandibular edentulism and cribra orbitalia (Mogle and Zias, 1995).

In order to tackle the problems behind the diagnosis of scurvy in past populations, Ortner carried out a series of studies examining the skulls of non-adults from Peru and North America (Ortner and Ericksen, 1997; Ortner *et al.*, 1999, 2001). The resulting series of cranial lesions are now considered highly indicative of the disease in young children (Table 6.2). These lesions are in the form of bilateral porosity and new bone formation, most commonly on the orbital roof, maxilla and the greater wing of the sphenoid, due to haemorrhaging of

Table 6.2 *Sites of pitting and new bone formation associated with scurvy*

Cranium	Mandible	Postcranium
Cranial vault	Medial coronoid process	Supraspinous process of scapula
Greater wing of sphenoid		Infraspinous process of scapula
Orbital roof		Metaphyses of long bones
Orbital aspect of zygomatic bone		
Posterior aspect of maxilla		
Internal aspect of zygomatic bone		
Infraorbital foramen		
Palate		

Source: Ortner *et al.* (2001:346).

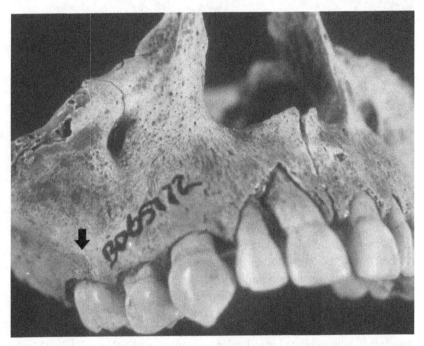

Figure 6.10 New bone formation around erupting second maxillary deciduous molar of a 3–4-year-old child from Beckford, Worcestershire, UK (100 BC–AD 100). (Photograph: Jean Brown.)

the anterior deep temporal artery underlying the temporalis muscle (Ortner *et al.*, 1999). The porosity is thought to occur as extravasated blood escaping from venous rupture stimulates a vascular response, and the formation of capillaries at the site (Fig. 6.10). In addition, although anaemia is a feature of the disease, porosity of the orbital roof (cribra orbitalia) may represent haemorrhage

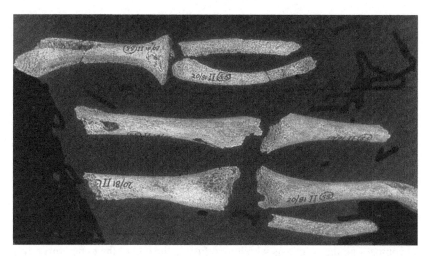

Figure 6.11   Infant with possible rickets and scurvy (20/81 II (55) no date). Note friable nature of the new bone formation and bowing of the radius, ulna and right tibia. (Photograph: Jean Brown.)

rather than marrow hyperplasia (Ortner *et al.*, 1999). Ossifying haematomas, or Parrot's swellings, may form on the parietals, occipital and, in some cases, the frontal bones and mimic porotic hyperostosis (Ortner *et al.*, 2001; Melikian and Waldron, 2003). Some endocranial lesions may also be the result of slow haemorrhage of the dura as a result of scurvy (Lewis, 2004). A study of 557 non-adult skeletons from all over North America by Ortner and colleagues (2001) suggests that cranial lesions occur most frequently in children aged between 3 and 7 years, perhaps as a result of rapid growth in these age groups.

Postcranial lesions have received much less attention in the drive to identify the condition, and in many cases, the more subtle lesions associated with scurvy may be masked or indistinguishable from the lesions caused by accompanying rickets (Fig. 6.11). Clinical texts do outline the type of radiographic features we should expect. Ortner and colleagues (2001) suggest that porous lesions that extend 10 mm past the growing metaphyseal end of the diaphyses are indicative of scurvy. Normally, these vascular channels would be filled in by osteoblasts as the bone grew. Radiographically, at the metaphyseal end of the bone, cartilage cells become reduced in size and number and are thickened by continuous layers of osteoid. This heavily calcified band is weak and made up of irregular trabeculae which are susceptible to fracture. Spurs result from the extension of this heavily calcified layer and periosteal deposition. Directly preceding this feature is an area of thinned trabeculae, appearing as a radiolucent band or 'scurvy line' on radiographs (Caffey, 1978). Once healing begins, this radiolucent scurvy line will disappear leaving a band of dense calcification, recognised only as a Harris line.

Subperiosteal haemorrhage and subsequent new bone formation is most frequently seen at the ends of the femur, tibia and humerus, but may be present on the entire length of the diaphysis. The new bone formed as a result of haemorrhage remodels resulting in a slight thickening of the affected bones (Caffey, 1978). A similar process is seen in the flat bones of the scapulae and ilia (Ortner, 2003). Ribs may fracture at the costal ends producing a 'rosary' similar to that of rickets. In the epiphyses, a shell of heavily calcified bone (Wimberger's ring) surrounds a rarefied ossification centre. Radiographs of the epiphyses can aid in the diagnosis of scurvy, particularly when the condition has healed, as the changes in the epiphyses are often still present some years after the other signs of the condition have been remodelled (Caffey, 1978). In some cases, the distal femur may become cupped due to fractures of the undermined metaphyses, encompassing the epiphysis which may fuse to the end of the bone preventing any further longitudinal growth (Caffey, 1978), although usually deformities in the surviving child are not evident (Ortner, 2003).

6.8   Summary

Weaning stress is often considered the culprit when there is a peak in mortality and retarded growth profiles at around the age of 2 years in a non-adult sample. Due to their association with growth disruption, linear enamel hypoplasias and Harris lines are often measured in adult remains to identify the age at which disruption occurred. But adult remains only provide an indirect measure of childhood stress. Measurements of nutritional stress in non-adult remains provide direct evidence on their state of health, and avoid the problems of bone remodelling and dental wear that may eliminate such signs in the adults. Recent advances in stable isotope analysis has begun a new era in the examination of weaning stress in past populations, and it is now possible to identify more precisely not only the age at which weaning occurred, but also what type of diet the child was weaned onto. Improvements in our ageing non-adult techniques, microscopic imaging of enamel defects and refinement of diagnostic criteria for rickets and scurvy have enabled us to provide a much more detailed picture of the interaction between childhood nutrition, infection and environmental change in past societies.

7 Non-adult skeletal pathology

7.1   Introduction

There are many pathological conditions commonly recorded on the adult skeleton that can be identified on non-adult remains. Dental disease (Lunt, 1972; O'Sullivan *et al.*, 1989, 1992; Watt *et al.*, 1997), specific infections such as leprosy (Lewis, 2002b) and syphilis, rheumatoid arthritis (Rothschild *et al.*, 1997), neoplasms (Lagier *et al.*, 1987; Nerlich and Zink, 1995; Barnes and Ortner, 1997; Alt *et al.*, 2002; Anderson, 2002) and congenital conditions including hydrocephalus, Binder syndrome and achondroplastic dwarfism (Huizinga, 1982; Black and Scheuer, 1996; East and Buikstra, 2001; Tillier *et al.*, 2001; Mulhern, 2002) have all been previously reported. In order for lesions produced by these conditions to be seen on the skeleton, the individual has to be immunologically compromised sufficently to develop the condition, but strong enough to survive the disease into its chronic stages (Ortner, 1991). Acute infections, such as the plague, whooping cough, smallpox, measles and scarlet fever, were known to be major causes of child death in the past, but often killed the individual before any skeletal lesions could develop. The chronicity of a disease depends on host immunity, and the number and pathogenicity of the invading organisms (Ortner, 2003). Therefore, not all childhood diseases will be evident in the skeletal record, and preservation, growth and the nature of paediatric bone can both aid and limit the diagnosis of diseases in non-adult skeletal remains. For example, the rapid turnover of bone in a child means that certain conditions such as rickets become more apparent as large quantities of structurally inferior new bone quickly replace the previously ossified cortex (see Chapter 6). In an adult, the disease is less evident as bone turnover occurs at a slower rate. Once the condition has resolved, however, accelerated growth of the child allows the inferior bone to be quickly replaced by normal tissue, causing both the macroscopic and radiographic signs of the disease to disappear from the skeleton within months (Harris, 1933).

Both direct and indirect evidence for childhood diseases can be retrieved from the archaeological record. The presence of fetuses within, or expelled from, the abdominal cavity of an adult female can provide indirect evidence of obstetric health hazards that affect both the mother and the child, such as obstructed labour, infection or haemorrhage. The child may have been badly positioned in the womb or there may be pelvic obstruction, for example if the mother has deformities associated with vitamin D deficiency. Congenital syphilis and tuberculosis, as well as more acute conditions, can induce spontaneous abortion and may account for some of the fetal remains in the archaeological record. Shortened limbs or an oversized infected bone in an adult will suggest an onset of the condition in childhood when the growth plate was still open (Lewis, 2000). The scope of this chapter cannot reasonably cover the vast array of pathologies that may affect the child, and many are no different in their development and appearance than in adult remains (e.g. dental disease, congenital conditions). Instead, this chapter will focus on conditions most commonly recorded on child skeletons, and on what the presence of these diseases may tell us about the social and environmental factors that contributed to their appearance in the past. The chapter ends with a section on the potential of skeletal and dental pathology in the personal identification of children in forensic anthropology.

7.2   Non-specific infections

In palaeopathology, non-specific infections are those for which a precise pathogen or cause (aetiology) is unknown. This category of lesions includes periostitis, osteomyelitis and osteitis (or non-suppurative osteomyelitis). In archaeological cases, when the lesions affect single bones, or have no particular distribution, it is difficult to identify the actual cause of the infection. Nevertheless, all three processes can occur in particular areas of the skeleton as part of a specific disease complex (e.g. treponemal disease, leprosy or tuberculosis). Understanding the mechanisms, cause and patterning behind the development of these lesions in adults and children is one of the foundations of palaeopathological study.

7.2.1   *Periostitis*

Periosteal new bone formation (periostitis) is recognised as the deposition of a new layer of bone under an inflamed periosteum as a result of injury or infection. The periosteum is a fibrous sheath that surrounds all the bones of the skeleton, with the exception of the endocranial surface of the skull and the joints. The sheath has two layers, with the outer layer consisting of a white fibrous tissue with a few fat cells, and the inner layer being made up of a dense network of fine elastic fibres (Williams and Warwick, 1980). This inner layer retains its osteogenic capacity throughout life and is vulnerable to traumatic separation and haemorrhage. The periosteum is bound to the cortex by Sharpey's fibres, which are less numerous and shorter in children, leaving the periosteum susceptible to rupture (Caffey, 1978). The invasion of a foreign organism (commonly *Staphylococcus aureus* or *Streptococcus*) causes inflammation, and the periosteum may become involved as a result of the direct extension of a nearby infection, or haematogenous spread of bacilli from a distant site (Ortner, 2003). The inflamed periosteum stimulates the osteoblasts to deposit osteoid on the cortical surface of the adjacent bone. Initially, the bone deposited is disorganised and has a porous appearance referred to as 'fibre' (or woven) bone, representing an active phase of infection. Later, the new bone layer becomes remodelled and organised with a system of Haversian canals; this smooth 'lamellar' bone is continuous with the original cortex, and its presence is diagnostic of an infection that occurred and healed well before the person's death. A mix of fibre and lamellar bone is indicative of a chronic, active infection.

The diagnosis of periostitis in non-adult skeletal remains is problematic. In the long bones, appositional (normal) growth involves the deposition of immature disorganised bone on the cortical surface. This new bone is macroscopically identical to the fibre bone deposited during an infection, or after trauma. When attempting to diagnose pathology based on radiographs of non-adult bones, Gleser (1949) stressed that features such as double contours, cupping and spurring, characteristic of periostitis, rickets and metaphyseal fractures, may also occur as the result of body positioning during radiography. For clinicians, movement of the child can cause difficulties for accurate imaging. In post-mortem cases, the incomplete removal of the soft tissues and growth plate may cause features that could be incorrectly identified as pathology once radiographed. Gleser (1949) also warned that increased formation and mineralisation of the long bones during the normal growth process may mimic pathological features in a 2–5-month-old infant, when signs of congenital syphilis, scurvy and rickets may be suspected. Concerned about the implications of this study, Shopfner (1966) examined the radiographic appearance of the long bones of 335 healthy premature and full-term infants, and noted periosteal new bone in 35% of cases. The new bone was invariably bilateral and affected the femur, humerus, tibia, ulna and radius in that order of frequency. While the new bone was concentric in most bones, the tibia was most commonly affected on the medial aspect. The bone deposits were thick, but not multilayered, and appeared on radiograph as double contours in the youngest individuals, before they became incorporated into the underlying cortex.

In the few archaeological studies that have identified non-adult periostitis, it is characterised as a unilateral, isolated patch of bone raised above the original cortex (Mensforth *et al.*, 1978; Anderson and Carter, 1994; Walker, 1997) (Fig. 7.1). In infant remains, difficulties in distinguishing pathology from growth

Figure 7.1   Raised periostitis in a 3–4-year-old child from Beckford, Worcestershire, UK (100 BC–AD 100). (Photograph: Jean Brown.)

has meant conditions such as birth trauma, child abuse, syphilis, rickets, scurvy, hypervitamintosis A and infantile cortical hyperostosis are rarely considered, despite their clinical frequency in newborns. Buckley (2000) examined the prevalence of subperiosteal new bone in a small sample of children less than 3 years old from pre-contact burial mounds in Tonga, West Polynesia, and noted 29% of bones displaying symmetrical new bone formation. Only one of the illustrated cases displayed isolated profuse deposits clearly distinguishable from normal growth. In a later study of the entire non-adult sample, the prevalence of new bone formation in the children was suggested as being indicative of yaws (Buckley and Tayles, 2003). East (2003) attempted to assess the prevalence of pathology in a small number of known-age perinates from Mexico and Tennessee, and warned that every individual examined exhibited some form of new bone on their remains. It may be possible to employ histological analysis to distinguish these normal deposits from those formed as part of a disease process, just as characteristic microscopic features such as 'polsters' and 'grenzstreifen' have been used to determine periostitis as the result of specific diseases in adults (Schultz, 2001). Without further investigation into the nature of the deposit, and a precise guide as to where and at what age bone growth occurs, we may never be able to identify the true extent of inflammatory episodes in a child's life.

New bone formation in the maxillary sinuses of a non-adult is indicative of chronic maxillary sinusitis as a result of inflammation of the mucous membrane. In children, rhinosinusitis as a result of upper respiratory tract infections is particularly prevalent. Clement and colleagues (1989) diagnosed nasal and sinus infections in 64% of children in Belgium under the age of 9, and today sinus infections are common in children attending nursery school, where close contact with a large number of children allows respiratory infections to be easily transmitted. Horiuchi and co-workers (1981) found the condition to be more prevalent in children living in areas of Japan with high levels of air pollution. In adults this condition has been found to be more prevalent in the urban centres of the past (Lewis *et al.*, 1995) although results are not universal (Panhuysen *et al.*, 1997). Maxillary sinusitis is identified in the archaeological record as deposits of new bone on the normally smooth surfaces of the maxillary antra (see Boocock *et al.*, 1995). The identification of an infection in non-adult sinuses is problematic, as before the age of 6 the antra are not fully developed and are difficult to visualise. Until the individual is around 15 years of age, the antra continue to develop, and pits are evident on the floor of the sinus as the antra grow to accommodate the developing permanent dentition, making lesions difficult to identify (Lewis *et al.*, 1995).

Just as new bone formation in the maxillary sinus denotes an inflammatory reaction, so new bone formation on the tiny auditory ossicles indicates middle-ear disease (otitis media). Otitis media is primarily a disease of children, usually occurring under the age of 4 years as the result of *Streptococcus pneumoniae* or *Haemophilus influenzae* infections (Aufderheide and Rodriguez-Martin, 1998). In chronic cases, swelling of the Eustachian tube and pus build-up can drain through the tympanic membrane (ear drum) causing periostitis, lytic lesions and even fusion (otosclerosis) of the stapes, incus and/or malleus. The infection may heal spontaneously or continue to recur into adulthood. One of the most serious complications of this infection is the tracking of pus from the middle ear internally to the brain, resulting in the death of the child. On some occasions, the infection may track externally into the air cells of the mastoid process resulting in mastoiditis (Roberts and Manchester, 2005). Today, nearly 80% of all children experience some form of otitis media before the age of 6 years, making acute and suppurative otitis media some of the most common childhood conditions in the modern world, with recurrent infections resulting in tinnitus, hearing loss and meningitis. Likewise, mastoiditis was reported in 20% of children in the pre-antibiotic era (Kipel, 2003).

Archaeological evidence for otitis media and mastoiditis has been mainly confined to adult skeletal remains (Birkby and Gregg, 1975; Bruintjes, 1990; Dalby *et al.*, 1993). Ear ossicles are at their full size in newborns and, with the exception of the incus, do not remodel in life (Scheuer and Black, 2004:84, 92). Hence, any infection in non-adults will be just as evident as in adult skeletal remains. Diagnosis relies on the individual recovery of the tiny ear ossicles, or the use of an endoscope to examine them in situ. The most effective way of diagnosing mastoiditis is by radiographs of the mastoid process, noting changes in the pattern and size of the air cells, although this may be a challenge in children under the age of 7 years when the width and depth of the mastoid process is undergoing rapid change (Scheuer and Black, 2004:92).

7.2.2   Osteomyelitis

In children, an abundant blood supply in the red bone marrow and at the ends of the long bones is essential to maintain rapid growth in these areas, but often results in a widespread haematogenous infection into the bone tissue. Pathogens rely on a wealth of iron to achieve their full growth and replication potential (Ulijaszek, 1990), and non-adult bone provides the ideal environment for their survival. The firm attachment of the periosteum to the cortex in adults limits the spread of infection and new bone formation. In non-adults, the more loosely attached periosteum may be ripped from the entire length of the shaft resulting in enlargement (hypertrophy) of the affected bone (Fig. 7.2). The shredding of the periosteum causes the creation of a new sheath of bone or an involucrum. Bacteria within the medullary cavity results in death (or necrosis) of the original cortex, known as a sequestrum, which is then expelled through a draining sinus or cloaca. When all three features are present, this infection is known as osteomyelitis (Ortner, 2003). In more chronic cases, the infection may not produce a sequestrum, but instead spreads extensively through the bone inciting osteoblastic activity, expanding the contours of the bone and reducing the size of the medullary cavity. In these cases, the bone lesion is referred to as sclerosing osteomyelitis or osteitis (Steinbach, 1966).

Osteomyelitis can develop directly at the site of an open fracture or due to an expansion of infection from an overlying lesion. However, it is the indirect spread of infection through the bloodstream that is the most common and serious form of the disease in children. In the pre-antibiotic era, one study demonstrated haematogenous osteomyelitis in 7% of infants and 80% of children (Treuta, 1959). In most cases, the infection spreads into a single bone, but in 20% of

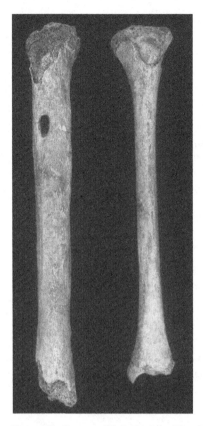

Figure 7.2 Osteomyelitis in the right tibia of an adolescent from Eccles, Kent, UK (sixth to eighth century AD). Note large cloaca, hypertrophy of the bone and elongation due to stimulation of the growth plate. (Photograph: Jean Brown.)

cases it will affect multiple bones (Fig. 7.3). The spread of the infection and which bones are affected is determined by the rate and timing of their growth, with the most common site for non-adult osteomyelitis at the knee (Ortner, 2003). Rasool (2001) reported that 71% of primary infections occurred in the tibial diaphysis in the 24 cases of childhood osteomyelitis that he examined. There is some debate about the protective quality the cartilaginous growth plate affords in the prevention of the spread of infection from the metaphysis to the epiphysis. When the epiphysis is involved, it is usually the result of a primary focus of infection, whereas secondary infections of the epiphysis due to a spread from the metaphysis are rare. In the fetal period and early infancy, the metaphysis and epiphysis still share the same blood supply making secondary and extensive spread of infection more likely (Shapiro, 2001). In addition,

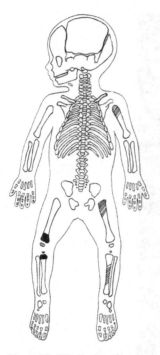

Figure 7.3   Distribution of haematogenous osteomyelitis in the non-adult skeleton. Black indicates most commonly affected areas; hatched lines indicate areas of less frequent involvement.

because in the shoulder and hip the metaphysis forms part of the articular surface, this allows for the spread of infection from the diaphysis into the joint capsule. Such a spread can have serious consequences for the developing child, with septic arthritis, necrosis, dislocation of the joint and shortening of the limb all potential complications (Ortner, 2003).

Due to the porous nature of the growing metaphysis in the child, osteomyelitis may be present without the formation of an involucrum or a readily identifiable cloaca, as pus is leached through the bone easily without subsequent vascular pressure and bone necrosis (Turlington, 1970). The stripping of the periosteum and subsequent new bone formation will remodel if the child survives the infection. In infants, haematogenous osteomyelitis could cause septicaemia and kill the child before chronic osteomyelitis can develop. This would account for the small number of child skeletons diagnosed with the condition in the archaeological record, compared to the high prevalence reported in the pre-antibiotic clinical literature. Ortner (2003) describes three cases in children aged between 6 and 9 years from different sites in North and Central America. In one case, the osteomyelitis is thought to have spread from a dental abscess. In two further cases, the hyperaemia caused by hypervascularity to the infected site resulted in massive elongation of the infected bone. This phenomenon will often lead to the premature closure of the epiphysis on the infected side (Caffey, 1978).

Osteomyelitis may be specific to fungal infections, tuberculosis, sickle-cell anaemia and congenital syphilis. It is a classic complication of chickenpox, and has a similar radiographic appearance to benign and malignant bone tumours (Schmit and Glorion, 2004). Typhoid fever (caused by *Salmonella typhi*) results in osteomyelitis in 1% of all cases, but it leaves no other distinguishable features, and cannot be diagnosed archaeologically. However, smallpox produces osteomyelitis (osteomyelitis variolosa) in around 20% of cases, and commonly affects multiple sites including the wrists, ankles and both of the elbows (Jackes, 1983). Ortner (2003:202) uses the presence of a Harris line in the unaffected bone to estimate that the child from La Oroya in Peru had suffered the condition for at least 4 years before its death aged 9. Prior to this, Canci and colleagues (1991) described a case of probable haematogenous osteomyelitis in a Bronze Age child from Toppo Daguzzo in Italy. All the surviving bones in the 5-year-old were affected, with the fibula showing the most extensive changes with several necrotic areas visible on radiograph. The authors concluded that the fibula was the most probable site of the initial infection, and that the child may have survived for up to 6 months after its spread. In England, Anderson and Carter (1995) reported a case of osteomyelitis in the hand (dactylitis) of a child from Canterbury in Kent.

7.3   Endocranial lesions

Reactive new bone, located on the endocranial surface in non-adults, is a relatively new area of investigation for childhood disease in the past. These features appear as diffuse or isolated layers of new bone on the original cortical surface, expanding around meningeal vessels; as 'hair-on-end' extensions of the diploë or, as 'capillary' impressions extending into the inner lamina of the cranium (Fig. 7.4). The lesions are commonly found on the occipital bone, outlining the cruciate eminence, but have also been recorded on the parietal and frontal bones, and appear to follow the areas of venous drainage. Various aetiologies have been suggested for these lesions in the literature including: chronic meningitis, trauma, anaemia, neoplasia, scurvy, rickets, venous drainage disorders and tuberculosis. All may cause inflammation and/or haemorrhage of the meningeal vessels (Griffith, 1919; Kreutz *et al.*, 1995; Schultz, 2001). Koganei (1912) was the first to describe endocranial lesions during autopsy, and identified pitted and grooved lesions, in addition to 'web-like' deposits on the internal surfaces

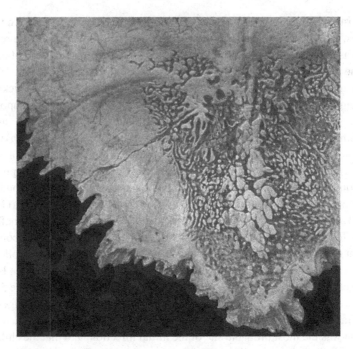

Figure 7.4 Hair-on-end endocranial lesions from St James and St Mary Magdalene, Chichester, Sussex, UK (AD 1200–1550). (Photograph: Jean Brown.)

of the frontal, parietal and occipital bones, which sometimes communicated with the diploë. Koganei found that lesions occurred more often in adults than children, and more commonly on the frontal than any other part of the skull. It was nearly 70 years before endocranial lesions were again described in non-adult remains. Mensforth and colleagues (1978) recorded healed and active vascularised 'periosteal reactions' on the endocranial surfaces in 645 of the Libben children, aged between 11 months and 2 years. They argued against the lesions being extensions of porotic hyperostosis from the ectocranial surface, but supported the view that they were the result of an inflammatory reaction.

Schultz (1989) found endocranial lesions in up to 22% of Bronze Age children from five sites in Central Europe and Anatolia, and recorded an increase in their frequency between the Bronze Age and medieval period. Schultz (2001) suggested that skull trauma resulting in an epidural haematoma and meningitis as a result of hydrocephalus were the cause of these lesions, and that the increase might be the result of population growth and socio-political change. In England, Brothwell and Browne (1994) noted endocranial lesions in eight children between 2 and 14 years of age from Jewbury in medieval York. They described the lesions as 'frosted' (remodelled bone projections) and 'vascularised' (a hypervascular inner lamina), but no aetiology was suggested, and the bones have since been reburied.

It is generally believed that the acute meningitis prevalent today would kill an individual too rapidly for bone lesions to occur. But children with the chronic form of meningitis in the early twentieth century were reported to survive for a month or even a year and often fell into a coma before they died (Griffith, 1919). Documentary evidence does suggest that children could survive for many weeks with a meningeal infection, long enough to produce new bone formation. Despite the increasing interest in these lesions in the literature, endocranial lesions are not considered in many core texts on palaeopathology, and their precise aetiology is unknown. Schultz (1984, 1989, 1993a, b, 1994, 2001) has recorded these lesions in samples of children throughout Europe, and histological analysis is often employed to try to understand the mechanism behind their appearance. This research suggests that hair-on-end lesions, described as 'calcified' plaques or 'pedicles', represent ossified soft tissue as the result of inflammation; that fibre bone deposits indicate active bleeding of the meninges, and that the vascular or capillary type of lesion is suggestive of healing. More recently, Hershkovitz and colleagues (2002) have suggested that respiratory diseases such as tuberculosis are the most likely cause of these lesions, although others have described the appearance of tuberculosis endocranial lesions to be isolated lytic 'corn-sized' impressions (Jankauskas and Schultz, 1995; Teschler-Nicola, 1997; Teschler-Nicola *et al.*, 1998). Morphological research on the occurrence of these lesions suggests that grey fibre bone deposits are most common in infant remains when the skull is undergoing rapid growth, and that the hair-on-end lesions, which extend beyond the cruciate eminence to the parietals and frontal bone, are more likely the result of a pathological condition (Lewis, 2004). In order to understand the aetiology of these lesions, researchers need to record the distribution, type and the age of individuals affected in addition to other skeletal lesions that may suggest a specific pathology. For instance, haemorrhage of torn meninges may be the result of child abuse and could be found in association with head trauma or rib fractures.

7.4   Infantile cortical hyperostosis

Infantile cortical hyperostosis (ICH) or Caffey's disease is an uncommon condition, and denotes a triad of lesions comprising swelling, irritability and bone lesions. The condition has both a familial and sporadic aetiology, heals spontaneously and has its onset and resolution in infancy. It is known throughout the world, affects all ancestral groups and has equal prevalence in boys and girls (Resnick and Niwayama, 1988:4118). The condition was most famously described by Caffey and Silverman (1945), and again by Smyth and colleagues (1946). Earlier reference to the condition was made in the German literature by Roske (1930), and it was alluded to by Caffey (1939) in a paper on infantile syphilis.

The specific aetiology of infantile cortical hyperostosis is still unknown. Modern cases are thought to result from a latent transplacental viral infection (Caffey, 1978), a genetic defect, hypervitamintosis, trauma, arterial abnormality or an allergic reaction of collagen. Clinically, the condition persists over a period of several months, and manifests as a localised tender lump on a rib or the mandible and, with less frequency, on the long-bone diaphyses (Aufderheide and Rodriguez-Martin, 1998:363). The average age of onset is 9 weeks, but the disease has been noted in utero (Barba and Freriks, 1953), and is rarely reported after 5 months (Caffey, 1978). But there have been active cases that have continued into the second or third year of life (Swerdloff *et al.*, 1970). The child may suffer from painful pseudoparalysis, pleurisy and anaemia (Resnick and Niwayama, 1988).

In the early stages of the condition, the periosteum becomes thickened and cellular, with the loss of the outer fibrous layer causing it to merge with the surrounding tissues. Osteoid is then deposited around the sheath and into the tissues. Profuse layers of new bone are formed causing the bone to appear thickened; while the medullary cavity retains its normal appearance. As the condition stabilises, the periosteum re-establishes its fibrous layer and the new layer of bone becomes incorporated into the original cortex. Remodelling begins from the endosteal surface, causing the medullary cavity to widen and the bones to become more fragile (Caffey, 1978). The skeletal distribution of the condition varies in the clinical literature, with early studies noting lesions on the long bones (Barba and Freriks, 1953), clavicle, scapula (Caffey and Silverman, 1945; Neuhauser, 1970), ribs (Caffey and Silverman, 1945; Barba and Freriks, 1953) and hands and feet (Caffey and Silverman, 1945), while later papers cite bone changes on the mandible as the most common (Swerdloff *et al.*, 1970; Blank, 1975; Finsterbush and Husseini, 1979). One clinical case reports bone changes to the skull thought to be indicative of ICH (Neuhauser, 1970). Currently, lesions are considered indicative when they appear on the mandible, clavicle and long bones (Couper *et al.*, 2001). All these lesions may be asymmetric and, unlike rickets and scurvy, the metaphyses and epiphyses are spared (Fig. 7.5). The condition can resolve within 3 months, but recurring forms of the disease may persist for many years and result in interosseous bridging of the radius and ulna, and between the ribs, dislocation of the radial head, mandibular asymmetry, bowing of the tibiae and other severe deformities (Barba and Freriks, 1953; Blank, 1975; Caffey, 1978; Resnick and Niwayama, 1988). Cases of recurrent

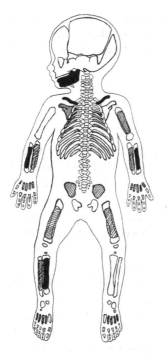

Figure 7.5 Distribution of infantile cortical hyperostosis (ICH) in the non-adult skeleton. Black indicates most commonly affected areas; hatched lines indicate areas of less frequent involvement.

ICH have been reported in individuals up to 19 years of age (Keipert and Campbell, 1970; Swerdloff *et al.*, 1970).

Archaeological cases of ICH are extremely rare. Rogers and Waldron (1988) reported two probable cases of ICH, one in a 10–18-month-old Romano-British infant, and the other in a 1-year-old from Anglo-Saxon England. Despite their poor preservation, the skeletons exhibited porotic hyperostosis, mandibular lesions and limited periostitis. Bagousse and Blondiaux (2001) identified profuse appositional new bone formation on the cranium and long bones of a child from Lisieux, France, and a staggering 57 cases of ICH were suspected at Roman Poundbury Camp, Dorset, perhaps due to recovery from smallpox (Farwell and Molleson, 1993). A reassessment of this sample suggests that the children were most probably suffering from rickets, scurvy or anaemia, with only one possible case of ICH identified (Lewis and Gowland, 2005). The confusion surrounding the aetiology and manifestation of ICH in the medical literature seems to be reflected in palaeopathology, with ICH used as a catch-all to describe widespread lesions on infant remains. Future research into the skeletal manifestations of the disease is needed. However, the non-fatal nature of the disease means that those children who may have developed the condition could recover completely, leaving no traces of the disease behind.

7.5   Tuberculosis

Tuberculosis (TB) is a chronic infectious disease caused by *Mycobacterium tuberculosis*. It can affect the lungs (pulmonary TB), lymph nodes (tuberculous adenitis), skin (scrofula), intestines (gastrointestinal TB) and, in some cases, the bones and joints. Tuberculosis usually has its onset in childhood and the identification of tuberculosis in non-adult remains is significant. Infected children represent the pool from which a large proportion of infected adults will arise and, because children are usually infected by adults, child cases indicate an ongoing transmission of TB within a community (Walls and Shingadia, 2004). Children are more likely to develop TB after exposure than adults, and modern estimates put the likelihood of infection after exposure at 5–10% for adults, 15% for adolescents, 24% for 1–5-year-olds and as high as 43% for infants (Walls and Shingadia, 2004). The disease will normally be evident between 1 and 6 months after infection, and in contemporary populations the ages between birth and 5 years, 15–30 years and 60 years plus are the age groups with the highest frequencies of the disease (Roberts and Buikstra, 2003).

Once infection is established, the primary lesion may lie dormant in a healthy individual after a brief inflammatory response, and may only become active later in life (secondary TB) when the immune status is compromised, or as a result of trauma to the affected bone or joint. In a more vulnerable individual, and especially in children, infection can lead to acute or miliary tuberculosis, where the bacteria spread to numerous sites via the bloodstream (Walls and Shingadia, 2004). Congenital tuberculosis is rare, but TB can spread to the developing fetus through the placenta, with the child ingesting infected amniotic fluid, through the birth canal, or via infected breastmilk (Roberts and Buikstra, 2003). Usually, children born to mothers with TB were stillborn or died shortly after birth (Holt, 1909). The most common mode of transmission is by inhalation of airborne droplets containing the bacilli from an infected person or animal, resulting in a primary infection in the lungs. In 1943, 61–67% of children dying from tuberculosis in London had the pulmonary form of the disease (Sheldon, 1943). The highest incidence of the disease was in children between the ages of 2 and 3 years, and the most susceptible had suffered from either whooping cough or measles, which were thought to lead to a reactivation of an old infection. A primary lesion in the lungs can spread within 6–8 weeks and affect the spleen, brain, joints and kidneys.

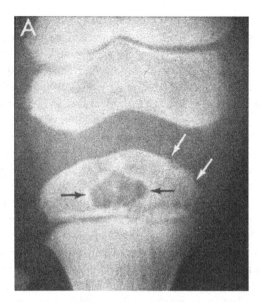

Figure 7.6 Tuberculosis in the epiphyses of a 3-year-old child. From Caffey (1978:673), reproduced with kind permission from Elsevier.

The haematogenous spread of TB usually results in the involvement of a joint, with the initial infection in the synovial fluid extending into the epiphysis and metaphysis (Fig. 7.6). Infection of the spine, hip and knee are most common, with the elbow also vulnerable. The joints may be affected bilaterally or unilaterally with equal frequency (Teklali *et al.*, 2003). Skeletal tuberculosis lesions are characterised by minimal bone formation (involucrum) and necrosis (sequestrum) and marked osteoporosis in the affected limb. Tuberculous granulomata destroy the joint cartilage and underlying bone, causing necrosis and pus formation, with the end result being fibrous or bony fusion (Shapiro, 2001). Although more rare, the bone shaft may become infected through a primary focus of infection in the nutrient canals, or through the infected metaphysis (Caffey, 1978). When this occurs, bone thickening under the inflamed periosteum is considered a common radiographic sign of the disease (Hayes, 1961). The tibia and femur are the long bones most commonly affected. In their study of the skeletal remains of children who died from tuberculosis in Portugal, Santos and Roberts (2001) also identified new bone formation on the shafts of the long bones, scapulae and tarsals. Stiffness of the joint, followed by skin abscesses, joint dislocation, muscle wasting and limb shortening are additional clinical symptoms of skeletal involvement (Teklali *et al.*, 2003).

The spread of *M. tuberculosis* to the meninges usually occurs during the primary infection and, is therefore, more common in young children than in adults

(Cappell and Anderson, 1971). This is a dreaded complication of the disease and, if the child survives, can cause blindness, deafness and mental retardation. Children between the ages of 0 and 4 years account for 80% of TB meningitis cases (Walls and Shingadia, 2004). In 1943, tuberculosis was the most frequent cause of meningitis following a pulmonary or gastrointestinal infection in England. It was common in the first 2 years of life, and children were reported to have survived for as long as a week before falling into a coma and dying 4–5 days later (Sheldon, 1943). In cases of miliary tuberculosis leading to meningitis in childhood, it is unlikely that the postcranial skeleton will show any signs of the disease, as this acute infection usually spreads haematogenously from a soft-tissue focus in the lung or intestine (Griffith, 1919). Hypertrophic osteoarthropathy may be evident in cases where this initial infection was overcome (Santos and Roberts, 2001), and vascular new bone may be located endocranially as a result of inflammation of the dura (Lewis, 2004).

Tuberculosis, or consumption, is a disease of overcrowding, be it in a thriving agricultural community or the unsanitary environments of rapid industrialisation. Poor housing and nutrition, close contact with infected animals and humans, poverty, climate and occupation have all been cited as causative factors (Roberts and Buikstra, 2003; Roberts and Manchester, 2005). The close proximity of infected individuals and a depressed immune status is required for the disease to take hold. It is not known when this disease first became a parasite of humans, but the human form of the disease is believed to have increased during the Neolithic period where it was spread from bovine TB with the widespread adoption of agriculture. More recent research suggests that TB was also carried by feral animals and was present in bison during the late Pleistocene; hence, it may have been present in human populations at a much earlier date (Roberts and Buikstra, 2003).

During the reign of Edward the Confessor (AD 1042–1066), the disease was known as the 'King's Evil'. Once a year the King would hold a ceremony where he would touch the tuberculous skin lesion (scrofula), and give a sovereign to the victim. By the sixteenth century, the disease was so widespread that Queen Elizabeth I stopped the ceremony claiming it was too costly. In the seventeenth century, during non-plague years, consumption was responsible for 20% of all deaths in England, and was the most common cause of death in the Western world (Roberts and Manchester, 2005). Signs and symptoms recognised as the King's Evil in medieval records could also have been caused by syphilis, arthritis, metabolic disorders and other lung infections, and it was not until 1689 that Morton separated scrofula into the tuberculous and non-tuberculous forms (Morton, 1689; cited in Chalke, 1962).

In the past, it was children who were most susceptible to the gastrointestinal form of the disease (Griffith, 1919). In children, gastrointestinal tuberculosis is usually fatal in the first 2 or 3 years of life, and results from the ingestion of the human or bovine strain of the bacillus. Gastrointestinal infection, with the human strain of the bacillus, may result from an ingestion of contaminated food, or may be secondary to pulmonary tuberculosis when infected sputum is swallowed (Sheldon, 1943). The bovine form of TB was transmitted through infected meat or milk, but may also have spread via the respiratory tract to people in close contact with their livestock. In gastrointestinal TB, the bacilli cause ulceration of the intestines and may spread via the lymphatic system to the mesenteric glands and the bowel, manifesting as a generalised infection or a localised abscess. The child develops diarrhoea, suffers abdominal pain, is anaemic and becomes emaciated as a result of malnutrition (Sheldon, 1943). One of the major causes of gastrointestinal TB in the past was the use of cows' milk for infant feeding. The desire for fresh liquid milk meant that milk started to be produced in urban cowsheds, or on the outskirts of towns. During the height of this practice (1850–60) reports of adulterated and infected milk were common. However, the transport of milk from the countryside could take up to 24 hours in unrefrigerated conditions before it was stored, uncovered, in shops or in the home (Atkins, 1992). By the 1870s, improvements in sanitation in the urban centres led to a decline in tuberculosis. It still remained a problem in the rural areas that did not benefit from the new reforms and, as late as 1931, even after strict rules had been applied to the storage and supply of milk, 6.7% of 'fresh' milk in England was still infected with bovine TB (Cronjé, 1984; Atkins, 1992).

Figure 7.7 illustrates the distribution of tuberculosis on the non-adult skeleton. One of the confounding factors in the study of tuberculosis in ancient populations is that not all people dying of the disease develop skeletal lesions. Skeletal tuberculosis only occurs in around 3–5% of people with the disease (Resnick and Niwayama, 1988). This figure may be higher in children whose blood-filled growing bones attract the bacilli. For example, in her study of the medical records of the Stannington children's sanatorium in Northumbria, UK, Bernard (2002) found the majority of the children were admitted for pulmonary tuberculosis, and that the bones and joints were affected in 12% of cases.

One of the most common lesions associated with pulmonary tuberculosis is new bone formation on the inner surface (visceral) of the ribs (Kelley and Micozzi, 1984; Pfeiffer, 1991; Roberts *et al.*, 1998; Santos and Roberts, 2001). These lesions are non-specific, and should not be considered pathognomic of TB as they have also been linked to other respiratory tract diseases such as pneumonia, actinomycosis and bronchitis (Lambert, 2002). In the spine, an infection may begin in the centre of the vertebral body and work its way into the intervertebral disk, resulting in collapse and kyphotic deformity (Pott's disease). Spinal involvement can lead to paralysis if the spinal cord is impinged (Lincoln and Sewell, 1963). In the pelvis, the acetabulum is more commonly affected,

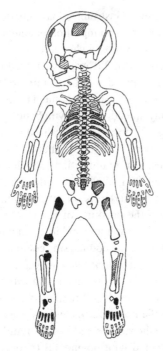

Figure 7.7 Distribution of tuberculosis in the non-adult skeleton. Black indicates most commonly affected areas; hatched lines indicate areas of less frequent involvement.

but the ilium may become involved due to the extension of a psoas abscess causing smooth cavitations. Due to the widespread haemopoietic marrow, and vascularisation at the growth plates, lesions of the hip, knee, ankle (especially the calcaneus) and tubular bones of the hands and feet (spina ventosa) are much more common in infants and children than in adults. In the phalanges, the entire shaft may be involved, becoming encased by an involucrum. If healing occurs, the digits may be shortened due to destruction of the growth plate (Ortner, 2003). Lesions of the ectocranium are also not common but, again, are more frequently found in children, affecting the cranial vault, cranial base and facial bones. The vault lesions appear as round foci up to 2 cm in diameter, and are greater in diameter on the endocranial surface. The lateral orbit, zygomatic, nasal bones, maxilla and mandible may also be involved and otitis media secondary to TB can result in destructive lesions on the petrous bone and mastoid process (Ortner, 2003).

Research into the identification of tuberculous meningitis by Lorber (1958), Teschler-Nicola and colleagues (1998), and Jankauskas (1999; Jankauskas and Schultz, 1995) suggest that lesions are likely in the form of small granulomas (cortical foci), producing lesions similar to arachnoid granulations. More recent research has begun to support this morphological characteristic. Hershkovitz and colleagues (2002) noted erosive lesions on the endocranial surfaces of skulls from the Hamann–Todd collection, and reported them in three out of 40 non-adults (7.5%). They found the lesions to be present in 36% of individuals documented to have suffered from TB, compared to 10% in those who died of other causes. Santos and Roberts (2001) carried out a study on 66 non-adults skeletons from twentieth-century Portugal, known to have suffered from tuberculosis during life. The second most common cause of death in these children was tuberculous meningitis, and three of the individuals were documented to have died from this condition. However, only 'slight' endocranial lesions were noted on the skeleton of a 19-year-old, non-tuberculous female with septicaemia.

Due to the vulnerability of children contracting TB in the past, the paucity of evidence for non-adults with skeletal lesions is not surprising. But several cases of childhood TB have been reported, with most being identified by the presence of Pott's disease (see Roberts and Buikstra, 2003). Of note are the studies now diagnosing TB in non-adults on the basis of aDNA. In Lithuania, Faerman and colleagues (1999) identified TB in a 15-year-old with no bone changes, and Pálfi and colleagues (2000) suggested tuberculous meningitis in an infant and a child aged 10–12 years in France, who displayed endocranial lesions and tested positive for TB ancient microbial DNA.

7.6   Congenital syphilis

One of the most controversial diseases identified in non-adult skeletons is congenital syphilis. The presence of the congenital form of syphilis in Old World cemeteries before 1493 would suggest that this venereal disease was endemic in Europe before the return of Columbus from the New World, and contributes to the debate raging between palaeopathologists surrounding the origin, evolution and transmission of treponemal disease (see Baker and Armelagos, 1988; Meyer *et al.*, 2002; Powell and Cook, 2005).

Congenital syphilis is a disease present at birth that develops in the fetus secondary to venereal syphilis in the mother. The causative organism, *Treponema pallidum*, can be transmitted to the fetus as early as the 9th week of gestation. In most cases, toxins released from dead micro-organisms invoke an allergic response and uterine contractions in the mother, resulting in fetal death and spontaneous abortion in the first half of the pregnancy (Genc and Ledger, 2005). These tiny skeletal remains, if they survive into the archaeological record, will show no signs of disease. At term a child may be stillborn, but in some cases a premature or sickly infant will be born alive. In these cases, the child may not display signs of the infection until it is about 2 years of age, or even until puberty (late congenital syphilis) (Ortner, 2003). In 1917, Harman reported pregnancies in infected women resulting in 17% spontaneous abortions, 23% stillbirths and 21% of the infants developing congenital syphilis. However, 39% of the group were unaffected at birth. The infection can also be spread to the newborn during its passage through the birth canal or due to contact with genital lesions, and although not spread through the breastmilk, lesions of the breast are a hazard to the sucking child (Genc and Ledger, 2005). Conversely, in the nineteenth century, it was noted that infants taken away from their mothers could infect the breast of the wet-nurse through the sores on their mouths (Lomax, 1979).

Today, congenital syphilis is rare in the West, with 0.02–4.5% of live births testing positive for antigens. Mothers carrying the infection are usually young, unmarried and less likely to seek the appropriate antenatal care and, hence, evade screening. An increase in use of crack cocaine has seen the incidence of congenital syphilis rise in minority groups in the USA. Congenital syphilis also increased in Russia in the 1990s with social changes as the result of the fall of the Iron Curtain, and it remains a problem in Africa with 30% of infant mortality in Zambia attributed to the disease (Genc and Ledger, 2005). Today, over a million pregnancies each year are adversely affected by syphilis (Walker and Walker, 2002).

As with other infections described in this chapter, the pathogen is transported haematogenously to sites all over the non-adult skeleton and, in particular, the rapidly growing knee (Fig. 7.8). There are two forms of bone and cartilage disruption or osteochondritis. In the 'passive form' the pathogen affects endochondral ossification as it occurs, while in the 'active form' syphilitic granulomatous tissue invades the bones and joints causing the bone to weaken and fracture, particularly at the growing metaphysis (Ortner, 2003). Periostitis usually develops during infancy, but it may be present in the fetus and, again, recognition of this lesion is currently hindered in the dry bone of infants. Genc and Ledger (2005) tabulated the clinical manifestations of congenital syphilis in early and late cases, with skeletal involvement being more common in the early manifestation of the disease, and dental and facial lesions more apparent in late onset congenital syphilis (Table 7.1).

The distribution of lesions is usually symmetrical and affects multiple bones, with circumferential profuse new bone formation, which may also involve the calvarium, mimicking porotic hyperostosis (Fig. 7.9). In older children, a mild infection during infancy will manifest as gummatous lesions (made up of syphilitic granulomatous tissue) characteristic of the adult form of the disease, but lesions will be less frequent and more contained. Caffey (1978) suggests that in milder cases of the disease, bone changes are less widespread

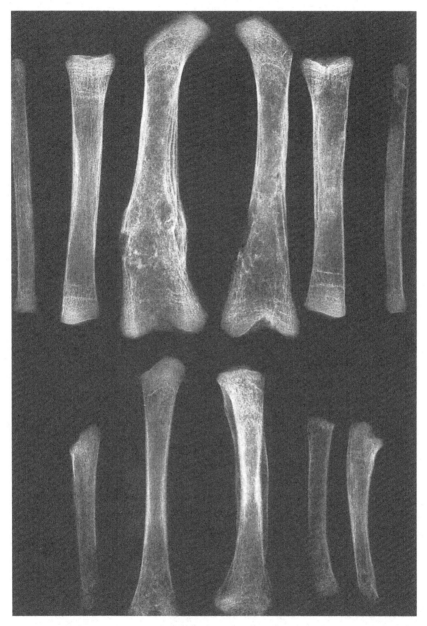

Figure 7.8 Radiograph of long bones of an infant from San Jeronimo's Church, Mexico City (17–18th Century) displaying fractures of the metaphyses in femora weakened by osteochondritis and infection. From Mansilla and Pijoan (1995:193), reproduced with kind permission from Wiley-Liss, Inc. a subsidiary of John Wiley & Sons, Inc.

Table 7.1 *Clinical manifestations of congenital syphilis*

Clinical findings	Percentage
Early onset	
Abnormal bone x-ray	61
Hepatomegaly	51
Splenomegaly	49
Petechiae	41
Skin lesions	35
Anaemia	34
Lymphadenopathy	32
Jaundice	30
Pseudoparalysis	28
Snuffles	23
Late onset	
Frontal bossing	30–87
Palatal deformation	76
Dental dystropathies[a]	55
Interstitial keratitis[a]	20–50
Abnormal bone x-ray	30–46
Nasal deformity (saddle nose)[a]	10–30
Eighth nerve deafness	3–4
Neurosyphilis	1–5
Joint disorder	1–3

[a]These three constitute the 'Hutchinson triad'.
Source: Genc and Ledger (2005:75).

and more commonly affect the metaphyses of the femur, tibia and humerus, and that the epiphyses are not involved in the disease. The bones most commonly affected in late onset congenital syphilis are the tibia, ulna and radius (Ortner, 2003). Flaring scapulae and medial clavicle thickening (Higomnakis' sign) may also be present (Powell and Cook, 2005). Radiographic features include diminished density of the metaphyseal ends and lytic destruction of the metaphyses due to the accumulation of granulomatous tissue, known as Wimberger's sign (Fig. 7.10). These lesions are most common on the medial aspects of the proximal tibiae and correspond to defects on the distal femora (Caffey, 1978). The 'sabre shin of Fournier' describes layered bone deposits on the anterior aspect of the tibia, while the medullary cavity retains its normal curvature (a feature that helps distinguish the macroscopic appearance of bowing from rickets on a radiograph). Caffey (1939) asserts that osteochondritis, periostitis and osteomyelitis are more indicative of syphilis when they occur together in the infant skeleton than when they occur independently. As with

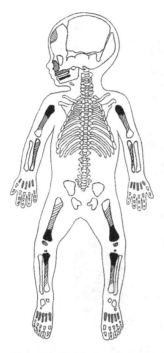

Figure 7.9 Distribution of congenital syphilis in the non-adult skeleton. Black indicates most commonly affected areas; hatched lines indicate areas of less frequent involvement.

tuberculosis, the abundance of red bone marrow in the fingers results in dactylitis in the child, and the development of a thin bone shell around the shaft. The proximal phalanges are most commonly affected on both hands, although the fingers affected may differ (Ortner, 2003). Another feature reported by early radiologists is the 'saw tooth' appearance of the metaphyseal ends of the long bones (Harris, 1933). These lesions have not been reported in osteological studies so far, and such jagged edges may be lost post-mortem, or may be mistaken for the 'bristles-of-a-brush' features more typically seen in rickets and scurvy.

In the skull, the developing gummatous lesions follow a different pattern to the sequential formation in adult caries sicca outlined by Hackett (1981). Facial changes are also more frequent. Destruction of the nasal cartilage and surrounding bone results in nasal collapse (saddle nose), and a respiratory infection develops, known as the 'syphilitic snuffles' in which: 'a profuse discharge of pus, perhaps bloodstained, [is] continually running from the nose, so that the infant cannot suck and is half asphyxiated if it tries to close its mouth' (Still, 1912:776). Although Still found milder degrees of this condition in 70% of his patients, he suggested that the more severe forms and the associated collapse

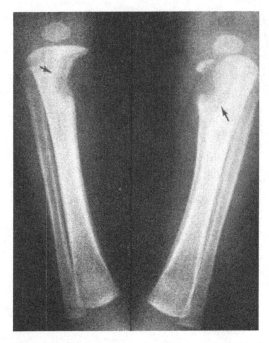

Figure 7.10 Wimberger's sign (syphilitic metaphysitis) in the proximal tibiae of a 2 month old child. From Caffey (1978:681), reproduced with kind permission from Elsevier.

of the nasal bridge were 'quite rare'. In congenital syphilis, the base of the skull may be particularly affected, causing hydrocephalus and distension of the scalp veins, in addition to spastic paralysis, mental defects, fits and pituitary disorders (Sheldon, 1943). In the rest of the skull, the palatal arch may appear abnormally high and the mandible and maxilla may be disproportionate in size (Powell and Cook, 2005).

At present, it is the dental stigmata of the disease that are considered pathognomic of congenital syphilis. These were most recently described and illustrated in detail by Hillson and colleagues (1998). Known as 'Hutchinson's incisors', 'Moon's molars' and 'mulberry molars', these dental defects occur in the early stages of congenital syphilis, perhaps in the first few months after birth, but they only become apparent with the eruption of the incisors and first molars around the age of 6 years. Defects of the incisors were first described by Hutchinson as being small, notched, easily worn and having dirty grey enamel (Hutchinson, 1857, 1858). These features were not evident on the deciduous teeth, and they may be hidden by a thin layer of enamel and/or dentine when they first erupt. The soft nature of the enamel means that these characteristic notches can be worn away (Fournier, 1884) (Fig. 7.11). Hillson and colleagues

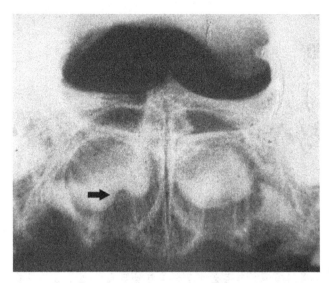

Figure 7.11 Radiograph of an infant from San Jeronimo's Church, Mexico City (17–18th Century) with Hutchinson's notches on the unerupted maxillary central incisors. From Mansilla and Pijoan (1995:193), reproduced with kind permission from Wiley-Liss, Inc. a subsidiary of John Wiley & Sons, Inc.

(1998) also report on defects in the permanent canines, where deep hypoplasias at the tip of the crown are broken off to leave a 'notch'. Moon's or bud molars describe the morphology of the first permanent molars characteristic of congenital syphilis, where the tooth is smaller than normal, and narrower at the cusp than the base of the crown (Moon, 1877). Mulberry or Fourier's molars are recognised as first molars with deep hypoplastic defects cutting into the cusps giving a 'berry' appearance to the tooth, but the secondary molars can also be affected (Hillson *et al.*, 1998). Although mulberry molars are not diagnostic of congenital syphilis, they are at least suggestive. Hutchinson's incisors and Moon's molars, however, are considered pathognomonic. In one study 45% and 22% respectively of children born with syphilis showed these features, compared to only 12% that showed skeletal evidence of the disease (Hillson *et al.*, 1998). Stodder (2005) argues strongly that postcranial lesions should be considered diagnostic of congenital syphilis in sites where dental stigmata are also present.

To date, 16 cases of possible congenital syphilis have been reported in the literature. Nine of these cases come from the New World and three pre-date Columbus' voyage to those shores (Table 7.2). Cook and Buikstra (1979) identified endocranial lesions in 27% of their Lower Illinois Valley non-adults and found an association between these lesions, periostitis, dental defects and a lower expectation of life. Cook had previously suggested meningitis secondary

Table 7.2 *Possible cases of congenital syphilis in the Old and New World*

Location	Date	Age (years)	Lesions	Reference
New World				
Virginia	925 BC	3 years	Reactive new bone on tibiae, severe enamel hypoplasias	Ortner (2003)
Virginia	Pre-Columbian	6–7 years	Reactive new bone on cranium, facial bones, long bones, dactylitis, severe deciduous enamel hypoplasias	Ortner (2003)
Illinois Valley	AD 200–1100	various	Endocranial lesions, severe dental defects, periostitis	Cook (1994); Cook and Buikstra (1979)
Vicksberg, Mississippi	AD 1322–1488	adolescent	Dentition only. Hutchinson's incisors, mulberry molars	Cook (1994)
Hawikku, New Mexico	AD 1425–1680	8–9 years	Bilateral Hutchinson's incisors, additional hypoplastic dental defects. No postcranium	Stodder (2005)
		15–19 years	Mulberry molars	
Newton Plantation, Barbados	AD 1660–1820	3 young females	Moon's molars, Hutchinson's incisors, 'pumpkin seed' canines	Jacobi *et al.* (1992)
Newton Plantation, Barbados	AD 1660–1820	5–6 years	Mulberry molars, notched canine. Isolated mandible, no cranium or postcranium	Curtin (2005)
Mexico City	17–18th century	2 years	Bilateral osteochondritis, osteomyelitis, hydrocephaly, rib periostitis, Hutchinson's incisors (unerupted), Wimberger's sign, Harris lines	Mansilla and Pijoan (1995)
Old World				
Metaponto, Italy	580–250 BC	Infant	Periostitis on femur and humerus, no dental defects	Henneberg and Henneberg (1994)
		3–4 years	Periostitis on distal femur, no dental defects	
		15–19 years	Mulberry molars	
Slaboszewo, Poland	14–19th century	12–15 years	Porotic hyperostosis, lytic lesions on frontal, gummatous lesions on clavicle, distal humerus and scapula, periostitis on radius and ulna, no dental defects	Gladykowska-Rzeczycka and Krenz (1995)
Szentkirály, Hungary	17th century	9–10 years	Atrophy of nasal spine, sabre shin, no dental defects reported	Ferencz and Józsa (1992)

to endemic treponematosis as a causative factor for the cranial and long bone reactive lesions (Cook, 1994). Pálfi and colleagues (1992) reported on an elderly (*c*.50 years) female skeleton and associated 7-month fetus recovered from the necropolis of Costebelle, France (Pálfi *et al.*, 1992). Despite the age of the mother, it was the fetus that showed pathological lesions with extensive new bone formation on the long bones, maxilla, ribs and cranial vault, and ankylosis of the hand and foot bones due to calcification of soft tissue. There were also possible destructive lesions on the parietal bones and Wimberger's sign on a radiograph. Pálfi and colleagues argued that the mother had been in the early stages of syphilis, perhaps only one year into the infection, where clinical evidence suggests nearly all pregnancies will involve the spread of infection to the developing child. The importance of this diagnosis was that it provided evidence for venereal syphilis in the Mediterranean basin from the first century AD, over 1390 years before Columbus returned from the Americas.

In order to try and evaluate our ability to diagnose congenital syphilis in the skeletal record, Rothschild and Rothschild (1997) examined 151 non-adults from the Buffalo Site in West Virginia, where clear signs of venereal syphilis were evident in the adults. They concluded that only 5% of children with treponemal disease will show evidence of dental or osseous involvement (a much lower figure than that given by Hillson and colleagues in 1998). They argue that periosteal new bone formation alone will not distinguish congenital syphilis from the other treponemal diseases acquired during childhood (e.g. yaws and bejel), that it is rare in congenital syphilis, and that when it does occur it is short-lived and remodels quickly to leave no visible trace on the skeleton. Moving their focus away from congenital forms of the disease alone, Cook and Powell (2005) reported 141 (5.1%) cases of treponemal disease in non-adults (0–20 years). The authors acknowledged that the lesions used to identify the disease were less specific than those used in adult cases. Cranial vault and nasal lesions were rare with the majority of children diagnosed on the basis of periosteal lesions or osteitis alone, which may not represent actual cases of the disease (Cook, 2005). What has become clear is that treponemal aDNA cannot be detected in suspected skeletons. Of the 27 skeletons analysed by Bouwman and Brown (2005) with lesions indicative of treponemal disease, 13 had evidence for mtDNA survival, but none of the skeletons had positive results when tested for treponemal DNA.

7.7   Skeletal pathology and personal identification

Following a biological profile based on the age, sex, ancestry (Chapter 3) and stature (Chapter 4) of a deceased child, the forensic anthropologist will endeavour to provide a positive personal identification (name) for the remains, based on evidence for dental treatment, previous injuries and any distinguishing characteristics the individual may display. The personal identification of non-adult remains provides many challenges, as young children rarely visit the dentist or receive extensive dental treatment or surgery, and they may not have been exposed to ante-mortem trauma or infection during their short lives. Nevertheless, certain pathological features evident on the bones and teeth have been explored as having the potential to provide information on the identity of the unknown child, and they warrant a mention here.

In forensic and archaeological science, the presence or absence of the neonatal line in the deciduous dentition has been used to assess whether a child was stillborn, or was possibly the victim of infanticide (Chapter 5). The neonatal line was first described by Rushton in 1933, and represents the first layer of enamel laid down after birth. The neonatal line is potentially present in all deciduous teeth and the cusps of the first permanent molars, and represents a hypomineralised area that undergoes a marked change in the direction of the enamel rods at birth. It is the difference between the partially mineralised enamel laid down before birth, and the fully mineralised enamel laid down after birth that characterises the neonatal line on thin sections. In some cases, the neonatal line may even show up on the tooth surface as an enamel hypoplasia (Noren, 1983). Eli and colleagues (1989) examined the average width of the neonatal line and reported that it was wider in children born by a difficult delivery and thinner in those born by Caesarean section. Skinner and Dupras (1993) explored the possibility of using the location of the neonatal line to reconstruct the birth histories of unidentified children, in the hope that it would provide a means of personal identification by suggesting whether the child was born preterm, fullterm or was overdue. Unfortunately, they showed that the position of the neonatal line in children born in the normal birth period (38–42 weeks) could vary by up to 9 weeks from its average position in the dental enamel, and that only the lateral mandibular and central incisors were reliable. Skinner and Dupras (1993) warned that the location of the neonatal line in normal-term births could vary so widely that it was unlikely to provide a reliable indicator of children who had non-term births. More recent studies have found a good correspondence between gestational age and the position of the line in infants (Teivens *et al.*, 1996; Smith and Avishai, 2005). Despite previous concerns, the neonatal line has become a popular marker in archaeological studies that examine the timing of stress events or age of weaning using counts of incremental lines (see Fitzgerald and Saunders, 2005). These methods have yet to be employed in forensic cases.

Where neonatal hair survives in association with skeletal remains, information on maternal habits such as smoking (nicotine) and cocaine (benzoylecgonine) abuse may be obtained. Levels of cotinine have even been used to suggest whether the mother actively smoked, or was a passive smoker (Koren, 1995). Children exposed to cocaine have lower birth weights and birth lengths than children of non-cocaine abusers and often need resuscitation and medical support at birth. Hence, the isolation of these chemicals in neonatal hair may help identify the parents of an unidentified child, or establish cause and manner of death (Koren, 1995).

Dental defects formed during childhood illness may aid in the identification of older children. For example, Skinner and Anderson (1991) published a remarkable case of child identification based on the timing of enamel hypoplasias in an unknown child's dentition. The child was a 6-year-old Native American from British Columbia in Canada. The age, time elapsed since death and location of the body all pointed to a particular child (presumptive identification), but the dental and medical history was absent, making a positive identification impossible. Skinner and Anderson (1991) examined the location of the striae of Retzius indicating episodes of pathology or stress in the child's life. When the medical records were eventually obtained, the striae corresponded to major incidents in his life such as entering a foster home, the death of his mother, surgery and hospitalisation from the combined effects of food poisoning, pneumonia and asthma, providing a probable, but not conclusive, identification.

Evans (2002) examined the effects of sports in children and their skeletal signatures. Intense physical exertion at a young age can lead to fractures in certain bones, and osteochondral lesions such as Osgood–Schlatter's disease, and osteochondritis dessicans (where a piece of bone and cartilage become detached from the joint surface). These conditions are now rare in the developed world, unless intense sporting activity is involved. For example, tennis players were shown to suffer traumatic injuries in their arms and ankles, as well as 'tennis elbow' in the form of osteochondritis dessicans. Rugby players suffered from fractures to almost all the bones of their body, with the exception of the pelvis and feet, and the lower legs were subject to trauma in ballet dancers. Ribs and skull injuries were expected in footballers (Evans, 2002). Bending and twisting of the spine by gymnasts and cricket bowlers can result in detachment of the neural arch in the fifth lumbar vertebra (spondylosis) and in the gymnast, micro-trauma to the radius growth plate has also been noted (Carty, 1998). Despite the obvious potential for future research, there are few cases in the published literature that have utilised the presence of skeletal or dental pathology in the identification of child remains, with one exception. In 2002, Dirkmaat published the case of two 3-year-old boys recovered from a housefire in Pennsylvania. They were distinguished from one another by matching post-mortem and ante-mortem radiographs of a lower leg oblique fracture suffered by one of the children (Dirkmaat, 2002). More cases of child identification need to be published in the forensic literature if we are to advance this important area of investigation.

7.8   Summary

There are many reasons why the skeletal distribution and manifestation of disease is different in children than in adults. For example, the widespread distribution of haemopoietic marrow in the child allows for greater dissemination of infections such as osteomyelitis and tuberculosis throughout the skeletal system. The porous nature of the growing bone means that abscess formation is often limited as vascular pressure is never built up (Turlington, 1970). Nevertheless, as the loosely attached periosteum is more easily stripped from the cortex, it can lead to more obvious and widespread involvement of the bone surface. Bone deformation is also more likely with infection of the epiphyses and joint capsule, which may stimulate abnormal growth, or destroy the growth plate causing fusion and crippling of the child.

A review of the childhood diseases that can be identified on non-adult skeletons suggests that as societies became more urbanised, respiratory infections such as sinusitis and tuberculosis increased. Non-specific infections (e.g. periostitis and endocranial lesions) are also useful measures of child health in the past and will become increasingly important as research into their causative factors progresses. It is clear that within our skeletal diagnosis, radiographs of the epiphyses should be included and may provide diagnostic criteria for tuberculosis, osteomyelits and syphilis. All show the importance of diagnosing diseases in non-adults to answer social questions about the past. Finally, pathological features on the skeleton and dentition of non-adults can provide valuable information when attempting to provide a personal identification for unidentified child remains in forensic cases.

8 *Trauma in the child*

8.1 Introduction

The identification of trauma in non-adult skeletons is rare compared to the rates recorded in adult samples. One of the main reasons for this is that fractures behave differently in children. It seems unlikely that children did not suffer injury in the past, but the nature of immature bone and rapid repair can mask the subtle changes, meaning that rates of non-adult trauma in the past are almost certainly an underestimate. Today, the most common forms of injury in the child are due to motor vehicle accidents, accidental falls (5–10-year-olds), intentional abuse (infants) and recreational sports injuries (adolescents). In the past, child's play, apprenticeships, warfare and physical abuse all exposed children to trauma and, although the mechanisms behind skeletal injuries may have changed over time, the nature of paediatric bone and its reaction to trauma has not.

8.2 Properties of paediatric trauma

The size of a child makes it vulnerable to serious injury in any collision; a moving vehicle (car, cart) will hit a small child in the chest or pelvis, as opposed to the lower legs of an adult. A child is lighter, and more likely to become a projectile, sustaining further injury when it hits the ground. The skeletal structure of a child will also influence the severity of injury. The highly cartilaginous and hence plastic nature of paediatric bone means it is less able to protect the vital organs before fracture. In addition, a child's ribs do not cover the liver and spleen, the small and large intestines are not protected, and the bladder is distended (Wilber and Thompson, 1998). A small child's disproportionately large and heavy cranium also means that during a fall from a height, the child is more likely to land on its head, as its shorter arms make it less able to protect itself by falling onto an outstretched hand (Wilber and Thompson, 1998). Hence, an injury to a child, from a fall or collision, is much more likely to cause fatal soft-tissue injuries, and peri-mortem fractures that are difficult to identify.

163

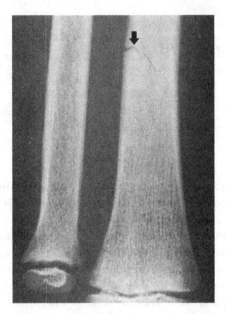

Figure 8.1 Greenstick fracture of the distal radius. Note that the fracture has not progressed through the entire width of the bone. From Resnick and Goergen 2002:2680), reproduced with kind permission from Elsevier.

The types of fracture seen in children are characteristic of paediatric bone's highly cartilaginous nature, areas of weakness and porosity. The highly plastic nature of non-adult bones commonly produces fractures that do not cause deformation and heal quickly, making them difficult to detect (Currey and Butler, 1975).

The most common forms of fracture in children are:

(1) greenstick fractures, with bowing of the compressed side and fracture of the tensile side;
(2) torus or buckle fractures, resulting in subtle bulge of the metaphyses as it fails under compression fractures; and
(3) plastic deformation, causing unusual bowing without fracture.

Currey and Butler (1975) carried out a series of experiments that showed greenstick fractures are common in children's bone due to the lower elastic but higher plastic threshold, causing paediatric bone to break more easily and with less force than mature bone. This strain load is not enough to force the fracture through the entire shaft (Fig. 8.1). The high plastic deformation threshold of children's bone means that the force is dissipated through transverse cleavage cracks, limiting its progression through the bone. In addition, the porous nature of the cortex means that some of the force is deflected from the surface of the bone, increasing the amount of pressure needed to cause the initial fracture (Currey and Butler, 1975). Nevertheless, porosity at the metaphyseal ends, together with a thinner cortical layer, makes this area of the bone particularly susceptible to trauma. The loose attachment of the periosteum to the cortical bone, and its thicker and stronger structure (Wilber and Thompson, 1998), will often result in a widespread haematoma and large callus formation along the shaft, despite limited evidence of a fracture line. That greenstick fractures are rarely recorded in non-adult archaeological remains may be due to the fact that, as in clinical medicine, these large sheaths of periosteal new bone are mistaken for scurvy, syphilis, osteomyelitis or bone tumours (Adams and Hamblen, 1991).

Plastic deformation resulting in traumatic bowing of the long bones in children can occur without evidence of a fracture line. Borden (1974) reported on eight clinical cases of plastic deformation of the radius and ulna in children aged between 28 months and 11 years, all of whom had fallen onto outstretched hands. The bowing was described as smooth and broad, with no sign of periosteal reaction that would indicate a greenstick fracture. In the long bones, fractures result from longitudinal compression forces that cause the bone to react in an elastic manner, bowing under the force. Once this force is removed, the bone returns to normal. If the force persists, the bone will develop plastic deformation, whereby bowing will persist, until eventually the bone fractures (Borden, 1974). Plastic deformation usually occurs in one of the paired bones (i.e. radius and ulna or tibia and fibula) but is most common in the ulna (Stuart-Macadam *et al.*, 1998). Cortical thickening is noted on the concave side of the deformity and there may be a reduction in pronation and supination of the forearm and angular deformity. In young children, rapid remodelling will correct the deformity, but Borden (1974) states that it may persist in children who sustain injuries after 10 years of age.

8.3   Types of fracture

The most obvious difference between adult and non-adult bone is the presence of the cartilaginous growth plate (or physis) at both ends of the long bones, and at one end of the short tubular bones. The growth plate is vulnerable to trauma that can result in avulsion fractures of the epiphysis away from the metaphysis (Adams and Hamblen, 1991). These physeal fractures may be complete or partial resulting in part of the metaphysis detaching from the diaphysis. Variations in the radiographic features of physeal fractures were first classified in 1963 by Salter and Harris (Fig. 8.2) and their system has since been expanded upon by

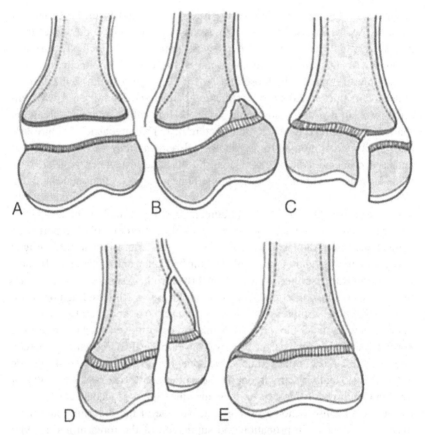

Figure 8.2   Salter-Harris scheme for fractures to the growth plate. From Resnick and
Goergen (2002:2723), reproduced with kind permission from Elsevier.

Ozonoff (1979) and Ogden (1981). Today, the most common areas for these
epiphyseal avulsion fractures in children are the distal humerus, proximal and
distal tibia and the distal femur (Adams and Hamblen, 1991).

Undisplaced epiphyseal fractures in infants and children are difficult to diagnose clinically, and usually require imaging of soft-tissue swelling to note
trauma and haemorrhage. These types of fracture are more common in infants
than older children, and often go unnoticed as the acute inflammation that causes
pain and immobilisation for a number of days in an adult can be resolved within
one day in a child (O'Connor and Cohen, 1987). In adults, displaced fractures
usually result in permanent deformity if not reduced. In children, rapid remodelling can cause any residual sign of deformity to be lost, and full strength and

function of the growth plate can return after just 10 days. Nevertheless, minimal, unilateral periosteal stripping and new bone formation may be evident for a brief period (O'Connor and Cohen, 1987).

8.3.1 Healing

Rapid healing in the child often causes fractures to go unnoticed, as fracture lines, callus formation and deformity are remodelled to retain the normal dimensions of the growing bone (Adams and Hamblen, 1991). There are three broad categories of fracture healing – cellular, metabolic and mechanical (Roberts and Manchester, 2005) – and the healing process of fractured bone is usually divided into six stages: (1) bleeding and haematoma formation, (2) haematoma organisation, (3) formation of the fibrous callus, (4) formation of the primary bone callus, (5) transformation of the primary (fibre) bone callus to a secondary (lamellar) bone callus and finally (6) the longest, mechanical stage; remodelling of the bony callus and realignment of the bone with eventual restoration of the normal architecture (Ortner, 2003:126; Roberts and Manchester, 2005:91). Only stages 3–6 will be evident on the skeleton, and the degree to which healing has occurred is most effectively assessed through anteroposterior and mediolateral radiographs. The time it takes for healing to occur, and the callus to be removed, depends on the type and severity of the fracture, the nutritional status of the individual, their age, alignment of fractured ends, the presence of infection or secondary pathological conditions and the type of bone fractured. For example, a radius takes less time to heal than a femur, and the upper limb bones normally take longer to heal than those of the lower limb (Roberts, 2000).

In general, although too variable to accurately predict, it is considered that in adults, cortical bone takes 3–5 months to heal depending on the size of the bone (Adams and Hamblen, 1991), and cancellous bone around 6 weeks (Ortner, 2003). The final stage, where the callus is finally removed, can take many years (Roberts and Manchester, 2005). The period of fracture healing in children, and especially infants, is greatly reduced (see Table 8.1). Salter (1980) illustrates the rapidity of the healing process in a child, related to age. If a femoral fracture occurs at birth, complete healing can take place within 3 weeks, in a child of 8 years this process could take 8 weeks, 12 weeks in a 12-year-old, and 20 weeks in a 20-year-old. In the abused child, constant trauma to an area will prevent bony callus formation, and may result in layers of periosteal new bone with neglect and malnutrition further delaying the healing process (O'Connor and Cohen, 1987).

Table 8.1 *Timetable of fracture healing (radiographic features) in children*

Category[a]	Early (e.g. infant)	Peak	Late
(1) Resolution of soft-tissue swelling	2–5 days	4–10 days	10–21 days
(2) Periosteal new bone	4–10 days	10–14 days	14–21 days
(3) Loss of fracture line definition	10–14 days	14–21 days	
(4) Soft callus (fibre bone)	10–14 days	14–21 days	
(5) Hard callus (lamellar bone)	14–21 days	21–42 days	42–90 days
(6) Callus remodelling	3 months	12 months	24 months to epiphyseal closure

[a]Repetitive injuries may prolong categories 1, 2, 5 and 6.
Source: After O'Conner and Cohen (1987:112).

8.3.2   Complications

Fracture complications in children are similar to those in adults, with vascular and neural injuries, fat embolisms, hypercalcaemia, malunion and non-union all cited in the clinical literature (Hensinger, 1998). In children, physeal injuries can result in ectopic bone formation leading to joint ankylosis. Premature fusion of the epiphyses will cause shortening, which will vary in its severity depending on the age of the child at the time of fracture, and the amount of longitudinal growth yet to be achieved. In cases where there is partial avulsion of the epiphyses, or only one of a pair of bones is affected (e.g. tibial epiphyses but fibular epiphysis intact), angulations of the joint surface may occur as the unaffected growth plate continues to develop normally, while growth of the affected bone is stunted (Adams and Hamblen, 1991).

8.4   Birth trauma

Birth injury is defined as any condition that adversely affects the fetus during delivery, and may result in brain damage as the result of hypoxia, or mechanical fractures (Gresham, 1975). Trauma may result from compression and traction forces during the birth process, abnormal intra-uterine position, difficult prolonged labour, large fetal size, and Caesarean sections. Significant trauma results in around 2% of neonatal deaths and stillbirths in the USA, with an average of 6–8 injuries per 1000 live births (Gresham, 1975). Typical birth injuries include fractures to the humeral midshaft and epiphyseal displacement of the proximal femur, as well as depressed fractures of the skull due to compression against the pelvic bones, abnormal positioning or excessive hand or forceps pressure (Brill and Winchester, 1987; Resnick and Goergen, 2002). A collection of blood between the skull and dura (cephalhaematoma) may develop with linear cranial fractures of the parietal and occipital bones during forceps use. In severe cases, the resulting haematoma can be the site for osteomyelitis or meningitis (Gresham, 1975). Spinal injuries resulting in death, transient or permanent paralysis occur most often in breech births (75%) where there is marked flexion and compression of the spinal cord, which may leave no bony evidence (Gresham, 1975). Clavicle fractures at the lateral and mid third of the shaft are very frequent birth injuries that often go unnoticed (Brill and Winchester, 1987). These occur when the shoulder is depressed during delivery, or comes into contact with the pelvis. The femur and humerus may be fractured in breech presentation, or as the result of uterine malformation (Resnick and Goergen, 2002). Fractures of the diaphysis in infants over 2 weeks old, with no evidence of callus formation, are more likely to result from abuse (Cumming, 1979). Neurological damage to the brachial plexus during childbirth can cause paralysis and atrophy of a limb (Erb's palsy). More rarely, plastic deformation due to faulty fetal positioning may occur (Dunn and Aponte, 1962), but may be confused with bowing due to rickets. In cases of osteogenesis imperfecta, severe and numerous fractures, dislocation and massive callus formation may be evident in the fetus simply due to pressures of the amniotic fluid, with more severe fractures occurring at birth. Such lesions were recently recorded in the skeletal remains of an infant with multiple 'crumpled' fractures of the long bones from a forensic mass grave in Guatemala (L. Rios, pers. comm.) (Fig. 8.3).

8.5   Non-adult trauma in the archaeological record

The study of trauma in past skeletal populations provides information on occupation, personal relationships, mortuary behaviour, accidents, subsistence and trauma treatment. As children were involved in many aspects of life within a community, and performed many subsistence and occupational activities, evidence for trauma in their remains helps to unravel questions such as the age of apprenticeship, child abuse, parental care, the home environment and, in the case of peri-mortem cuts during autopsy, the development of paediatrics. Despite the wealth of evidence for trauma in adult individuals in the archaeological record, the evidence for trauma in non-adults is very limited and is probably due to the difficulties in identifying these lesions in the child, as well as our failure even to examine children for such pathology in the skeletal sample.

In a review of hundreds of skeletal reports aimed at assessing past patterns of health and disease in Britain, Roberts and Cox (2003) did not find any recorded evidence of trauma in non-adult remains from the Neolithic through

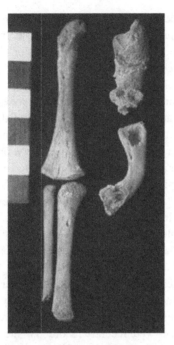

Figure 8.3 Bones of a child from a mass grave thought to have been suffering from osteogenesis imperfecta. Note the absence of the fibula, bowing of the tibia and fracture and deformity of the left femur. (Photograph: Luis Rios.)

to the Anglo-Saxon period, and non-adults did not even warrant a category in the discussion of trauma in the later and post-medieval chapters. In Norway, Fyllingen (2003) reported an absence of violent trauma in the children from a mass grave in Bronze Age Nord-Trøndelag, despite its high occurrence in the adults. That children were involved in warfare and massacres is evidenced by cranial trauma in infants and children from mass grave sites in Neolithic Germany (Whittle, 1996:170), the Palaeolithic Sudan and Mesolithic Bavaria (Thorpe, 2003). Dawson and colleagues (2003) reported three peri-mortem depressed fractures and chipped teeth in the skull of a 13–14-year-old child (reportedly male) from a pit in Chalcolithic Israel (4500–3200 BC). The authors argue the child received the blows during face-to-face combat, suggesting that in their teens, male children in this society became warriors. Equally, trepanations are rare in child skulls, with an early study of Peruvian skulls by MacCurdy (1923) showing no evidence of this treatment, although 17% (47 of 273) of the adults were affected. The difficulty in recognising postcranial fractures is illustrated by the subtlety of lesions such as those provided by Ortner (2003:148). Ortner described a 15-year-old child from Chichester, Sussex, UK with one

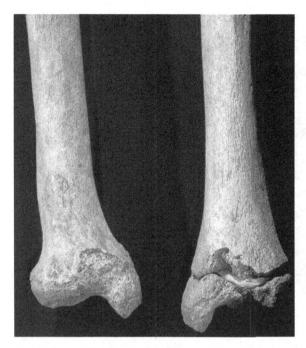

Figure 8.4   Fracture of the right distal tibia resulting in premature fusion of the growth plate in a 15-year-old from St James and St Mary Magdalene, Chichester, Sussex, UK. (Photograph: D. J. Ortner.)

fused and one open distal tibial epiphysis (Fig. 8.4). The child died before any tell-tale shortening of the affected leg could become evident. The second case was of a 9-year-old child from Pecos Pueblo in New Mexico, showing very subtle changes to the distal proximal metaphysic, with a callus mimicking periostitis, and no evidence of deformity (Fig. 8.5).

Fractures have been recorded in children from various periods in the past, and can reveal information about their treatment and activities within the society. In a study of four sites from medieval and post-medieval England, Lewis (1999) noted fractures in 1.5% ($n = 13$ individuals) of the non-adult sample. At Raunds Furnells, a 7–8-year-old child displayed bilateral fractures of the clavicle, and three cases of head injury (blunt force) were recorded in rural Wharram Percy. Here, one 5-year-old child had no evidence of healing suggesting a subdural haematoma and subsequent death. A shortened left leg in a child aged 6.5 years is suggestive of trauma affecting the growth plate. Trauma to the epiphyses was also recorded in the arm of one child, and the leg of another from post-medieval Christ Church Spitalfields in London, resulting in shortening of the limbs and in the latter, leaving the child with a limp. There was also evidence of healing

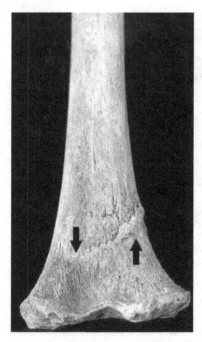

Figure 8.5 Probable stress fracture of the distal right femur in a 9-year-old child from Pecos Pueblo, New Mexico. (Photograph: D. J. Ortner.)

rib fractures in an 18-month-old child that, considering its age, may provide evidence of child abuse (see Section 8.6).

Stuart-Macadam and colleagues (1998) reported two possible cases of plastic deformation of the ulna in skeletons from Ontario (AD 1280–1550), one in a 12–17-year-old child. Later, an attempt to redress the balance in recording non-adult trauma was made by Glencross and Stuart-Macadam (2000) who outlined the clinical mechanisms of fractures and provided two case studies. They suggested that taphonomic factors of plastic deformation obscure many true fractures, and recommended radiographs to view angulation of these subtle changes. Glencross and Stuart-Macadam (2001) went on to provide two radiographic measurements: the humerotangential angle (HTA) and the anterior humeral line (AHL), as a method of diagnosing cases of childhood distal humeral fractures in adult skeletons. This technique involves collecting data for the mean male and female lateral and medial joint surface angles to identify those that fall outside of this mean. For the HTA, anteroposterior radiographs are used to draw a vertical line through the centre of the humeral shaft and the inferior joint surface, and a second line is drawn horizontally along the bottom of the joint surface. This line indicates the displacement of the joint surface on the transverse plane. Normally, the medial angle is larger than the lateral angle due to the trochlea. In healed trauma, the lateral angle may be 5°–10° greater and 5°–25° smaller than unaffected cases (Glencross and Stuart-Macadam, 2001). The AHL is constructed on mediolateral radiographs with a vertical line drawn along the anterior humeral cortex to the capitulum, and a perpendicular line drawn for the inferior to the posterior extent of the capitulum, and divided into thirds. In unaffected bones, the AHL passes through the middle third. True lateral radiographs are needed and can be achieved by ensuring superimposition of the posterior supracondylar ridges of the humerus on the radiograph. In one case, more subtle macroscopic evidence for cortical 'buckling' may be seen in the irregular morphology of the olecranon.

The most common evidence we have for childhood fractures in the past is the presence of shortened limbs in adults (Lewis, 2000). These result from the post-traumatic formation of a bony bridge between the metaphysis and epiphysis, causing premature fusion of the growth plate and shortening. How much shortening occurs depends on the age of the individual at the time of the trauma, the epiphysis affected and the amount of longitudinal growth still to occur (Fig. 8.6). For example, in the tibia, 43% of longitudinal growth from the centre of ossification is carried out by the distal growth plate, whereas 57% occurs at the proximal end of the bone (Maresh, 1955). Hence, a fracture and fusion of the proximal tibial epiphysis would cause a greater degree of shortening (growth dysplasia) than similar trauma to the distal end. In theory, detailed knowledge of proportional bone growth could provide evidence of the age at which the fracture occurred.

Evidence for birth trauma in past populations is equally scanty, but we may expect to see peri-mortem fractures in the skull of perinates from the medieval period and prior to 1726, when the Chamberlain family secret of birthing forceps was finally revealed (Rushton, 1991). Before this time it was the practice of midwives and man-midwives in England to employ a 'crochet' to extract difficult births from the mother where a successful live birth was deemed impossible. This implement comprised hooks that would be placed in the orbits and roof of the mouth of the child for extraction (Eccles, 1982). If the child was alive before the procedure, they would have certainly died during or shortly afterwards. The thin, fragile and fragmentary nature of infant bones may mean that these peri-mortem crushing injuries are missed and destroyed post-mortem. Loudon (1992) reports that in the eighteenth and nineteenth centuries, doctors in Britain were particularly notorious for favouring craniotomy over Caesarean section, a procedure in which the head of the child is removed in the womb to relieve the obstruction. Loudon estimated that four to six craniotomies were performed in every 1000 births compared to one in every 3000 in the rest of Europe. Again, the paucity of skeletal remains from these periods, and the difficulty in

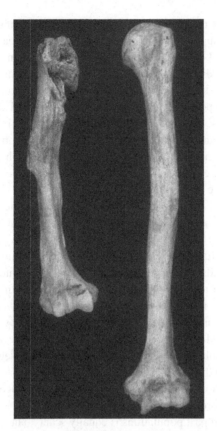

Figure 8.6   Shortening of the right humerus in an adult probably as the result of
trauma to the proximal growth plate in childhood. Skeleton from St James and St Mary
Magdalene, Chichester, Sussex, UK (AD 1200–1550). (Photograph: Jean Brown.)

identifying cut marks in thin or unfused bones, means that such procedures
may go unidentified in the archaeological record. Evidence of a birth injury to
the brachial plexus comes from Romano-British Kingsholm, Gloucester, UK,
where a young female exhibits evidence of Erb's palsy, causing wasting of the
right clavicle, scapula, humerus, radius and ulna, and in the right hand (Roberts,
1989). At Christ Church Spitalfields, a healed fractured clavicle was evident
in a 4-month-old child that may have been the result of birth trauma (Lewis,
1999). Molto (2000) describes two cases of humerus varus deformity in a male
and a female from a Roman cemetery in Dakhleh, Egypt which he interprets as
evidence of birth trauma to the proximal epiphysis.

   A more unusual case of a possible mechanical abortion or embryotomy was
reported from Romano-British Poundbury Camp, Dorchester by Molleson and

Cox (1988) suggesting that inhabitants had followed procedures laid out by the second-century physician Soranus of Ephesus (AD 200). The neonate was found within a coffin, with its head and left arm beneath the hips to the west end of the burial. The femur, right radius and ulna were above the left scapula, suggesting dismemberment at the time of burial and, perhaps, confusion in identifying the arm from the leg during deposition (Molleson and Cox, 1988). Peri-mortem cut marks were evident on the third cervical vertebra, the right humeral shaft, and the humeral head, which had nicked the right glenoid cavity. Cut marks were also recorded on the right femur, presumably to remove the tuberosity of the right ilium. The authors suggest that this child may have been very large, resulting in difficulties at birth.

In summary, birth trauma and obstetric intervention may be suspected if the following lesions are identified in the infant:

- Birth trauma: fractured clavicle
- Use of the crochet: peri-mortem sharp force trauma to the orbits and palatine surface of the maxilla
- Use of forceps: peri-mortem crush fractures to the parietals, frontal and/or occipital
- Craniotomy: cut marks on the cervical vertebrae
- Embryotomy: peri-mortem cut marks on the postcranium.

8.6   Physical child abuse

> Barn vil ikka teea, barn vil ikka teea,
> Tak an leggen, slog an veggen,
> Barn vil ikka teea.
>
> The child will not be quiet, the child will not be quiet,
> Take it by the leg and hit it against the wall,
> The child will not be quiet.
> Old Norse children's rhyme, seventeenth-century Shetland, cited in Knight (1986:136)

Today, 200 000 incidents of physical child abuse are reported in the USA per annum (Resnick and Goergen, 2002). Of these, 2500 children will die of injuries to the head or abdomen, and 25% will be under 2 years of age. The complete absence of any reported cases of child abuse in the archaeological record, from both Europe and North America, has led many anthropologists to suggest that child abuse, in its current form, is a recent phenomenon (Walker, 1997; Waldron, 2000). Documentary evidence would suggest otherwise, and the lack of data may be more the result of our inability to recognise it in non-adult skeletons. Soranus commented on the neglect and abuse of children by their wet-nurses (Waldron, 2000), and the Persian physician Rhazes (AD 860–923) also recognised that children could be victims of physical abuse and neglect (Lynch, 1985). Stone (1977) makes reference to a plea from medieval physicians against the common practice of vigorously rocking children to make them sleep when in fact they rendered the child unconscious, and in the Anglo-Saxon period there is reference to mistreatment of deaf and mute children (Kuefler, 1991). The Old Norse rhyme cited above suggests that physical abuse of children in the past was certainly not unheard of. This realisation caused De Mause (1974:40, 31) to state that: 'a very large percentage of the children born prior to the late eighteenth century . . .were . . . battered children . . . ' and that 'urges to mutilate, burn, freeze, drown, shake and throw the infant violently about were continuously acted out in the past'. While many of the acts of psychological and sexual abuse described by De Mause paint a grim picture of childhood in the past, they would be difficult, if not impossible, to identify in the skeletal record.

In 1860, the French physician Ambrose Tardieu acknowledged the existence and causes of child abuse in an article (Tardieu, 1860) that was largely ignored by his peers. For example in 1888, Samuel West reported several cases of periosteal swellings in young children from the same family, but did not suggest that abuse might have been a factor. Instead, West (1888) considered scurvy and syphilis as possible aetiologies before deciding acute rickets was the cause. In the 1930s there was an increasing recognition of battered children in baby hospitals in the USA that culminated in Caffey's influential paper in 1946. Here, Caffey (1946) discussed cases of subdural haematomas coupled with long-bone and rib fractures that he could not qualify with any known pathological syndrome. Although he noted that unreported trauma might be the cause, he hesitated in assigning these lesions to child abuse in the home, a conclusion that would have been highly controversial. This subject was taken up again by Silverman (1953) who suggested that periosteal swellings and metaphyseal fragmentation might be part of the same aetiology. Although he reports that these 'spontaneous' injuries only seemed to occur whenever the child was released from hospital, he warned that 'extreme care must be exercised not to overwhelm responsible custodians of the infants with guilt' (Silverman, 1953:426). Finally, Kempe and colleagues (1962) coined the phrase 'battered baby syndrome' to draw the public's attention to a phenomenon now widely recognised, if not acknowledged, by the medical community.

8.7   Clinical features of child abuse

Most of our knowledge about the lesions commonly associated with child abuse comes from the forensic medical literature, where forensic pathologists also rely on family circumstances, previous medical records, soft-tissue injuries and radiographic features that are not available to the anthropologist. In clinical medicine, the most common fractures associated with child abuse are of the ribs. In addition, spiral and metaphyseal fractures of the humerus, metaphyseal fractures of the femur, tibia and fractures of the scapula (acromion) and small bones of the hands and feet are also symptomatic (Brogdon, 1998; Resnick and Goergen, 2002). Fractures of the first rib and pelvis are considered to represent the most severe injuries in a child (Wilber and Thompson, 1998; Starling *et al.*, 2002). Reports on the frequency of fracture in child abuse cases range from 11% to 55% (Kleinman, 1987), and Wilber and Thompson (1998) reported injuries to the head and face in 40% of child abuse victims. In a survey of 165 postcranial fractures from suspected child abuse cases, Kleinman and colleagues (1995) examined 31 cases of child abuse and found fractures of the ribs (51%) and long bones (44%), with 89% involving the metaphyses, hands and feet (4%), spine (1%) and the clavicle (1%), although fractures of the clavicle, are considered extremely rare in child abuse cases (Brogdon, 1998). In forensic cases it is vital to be cautious as 'spontaneous fractures' of multiple long bones have been noted in autopsies of children suffering from cerebral palsy, and disuse atrophy can weaken bones that then fracture, often unnoticed, through day-to-day handling by the caregiver (Torwalt *et al.*, 2002). Brill and Winchester (1987) outlined a number of conditions that will mimic skeletal signs of child abuse including pathological fractures as the result of syphilis, osteomyelitis, rickets, scurvy, congenital indifference to pain, bone tumours and genetic disorders such as Menkes' syndrome (kinky-hair disease). Although many produce fractures of the metaphyses, some, such as infantile cortical hyperostosis, osteogenesis imperfecta and Menkes' syndrome, can produce fractures of the skull and face and must be ruled out when child abuse is suspected.

Multiple fractures of the skull and bilateral injuries that cross sutures are also indicative of abuse. In children, radiating linear fractures may breach the suture line and cause diastasis or opening of the sutures. This is most common at the sagittal suture, but may cause the metopic suture to reopen. In infants, a fracture may pass over the suture line into the next bone plate (most commonly the parietal), and may be 'stepped', rather than forming one continuous line (Knight, 1996). Pond lesions, characterised as shallow depressed fractures, are more common in the pliable bones of children, and may occur without fracture. The diagnosis of head trauma in the skeletal remains of children recovered from clandestine graves must be carried out with knowledge of the context in which the remains were found, and the type of weather to which the remains may have been exposed. A survey by Crist and colleagues (1997) demonstrated that lesions associated with head trauma in children, such as suture separation and brain swelling resulting in superior bone displacement, are also characteristic of exposure to sunlight, drying and humidity post-mortem. In burnt remains, the pattern of cracking related to the fire commonly resembles a spider's web, and affects the side of the skull, often bilaterally. Fractures to the base of the skull due to heat are virtually unheard of and must be regarded as ante-mortem injuries until proved otherwise (Bohnert *et al.*, 1997). In cases of gunshot wounds, even after reconstruction, the bones of the skull are usually so thin that it may still be difficult to be certain of the direction of the path of the projectile, or to determine the type involved (Di Maio, 1999).

In survivors of abuse, ossified subdural haematomas may be evident on the skeleton in the form of endocranial lesions of fibre bone, or as hair-on-end bony projections (Lewis, 2004). As blood accumulates within the subdural space, the brain is displaced adding increasing traction to the bridging veins, which haemorrhage with subsequent trauma (Kleinman, 1987). After the bleeding has ceased, the blood clot is surrounded by a fibrous, vascularised membrane, which later becomes detached from the dura, and usually extends bilaterally to the posterior parietal region, or to the floor and anterior/middle cranial fossa (Caffey, 1978). Blondiaux and colleagues (1998) reported successive plaque-like deposits on the endocranial surface of a non-adult skeleton from fourth-century Normandy, suspected of having suffered child abuse. The possibility of the severe rocking of a child accidentally resulting in subdural bleeds, as recorded in the medieval period, becomes suspicious in the light of modern experiments carried out in forensic cases (Jones *et al.*, 2003). The cause of subdural and retinal bleeding from shaken baby syndrome has recently come into question. There are suggestions that the physiology of the infant brain makes it susceptible to hypoxia and venous leakage, which should be considered as an underlying cause in cases where skull fractures are not evident (Geddes and Whitwell, 2004).

Characteristic radiographic signs of abuse in the clinical literature include corner and 'bucket-handle' osseous fragments of the metaphyses. Corner fragments result when a fragment of bone and cartilage are avulsed from the corner of the metaphysic. Bucket-handle deformities describe crescent-shaped fragments at the zone of provisional calcification between the metaphysis and epiphysis (see Salter–Harris scheme: Fig. 8.2). In addition, subtle areas of radiolucency may be evident where a portion of the metaphyses is left attached to the epiphysis and becomes sclerotic, emphasising the lucency of the provisional calcification zone (Kleinman, 1987). Another commonly reported feature of abuse in the clinical literature is periosteal new bone formation as a result of the rupture to the fragile periosteum, and subsequent haemorrhage. The distribution of these lesions is variable, and as the fibre bone applied to the subperiosteal surface is identical to that laid down during the neonatal growth process, there are concerns that misdiagnosis could occur (Plunkett and Plunkett, 2000). New bone formation on the lower limbs of a child, in combination with metaphyseal fractures, may occur as the result of the twisting and pulling of the limbs during a breach delivery (Snedecor *et al.*, 1935; Brill and Winchester, 1987). The characteristic distribution of periostitis in cases of shaken baby syndrome would be expected as the result of haemorrhage at the insertion points of the sternocleidomastoid, the ventral aspect of the scapula, underlying the brachial plexus at the proximal humerus, and on the proximal femur at the attachment of gluteus maximus (Saternus *et al.*, 2000). Diaphyseal new bone formation will be present where the child is grasped by the arms or legs when shaken, or lesions may be accompanied by posterior rib fractures when held and squeezed at the torso (Caffey, 1974; Kleinman and Schlesinger, 1997). Of course, for periostitis to be evident on the skeleton, the child would need to survive the incident for 4–14 days (O'Connor and Cohen, 1987). In cases of shaken baby syndrome where the child has survived, haemorrhage and dislocation of the cervical and upper thoracic vertebrae has been suggested to cause long-term spinal kyphosis (Cullen, 1975).

There have been several cases in which forensic anthropologists have examined skeletons of children who were known to have been victims of abuse in life. Kerley (1978) examined the remains of three young siblings who had been murdered by their parents. Their remains were exhumed when the surviving children mentioned in class that they had had other brothers and sisters, but that their parents had killed them! Multiple fractures were noted on the mandible, ribs, clavicles, radius and ulna. In 1997, Walker reported four cases in which anthropological analysis of remains recovered from clandestine graves revealed evidence of child abuse that had gone unrecognised by pathologists. In the first case, despite being fragmentary, the parietals displayed three episodes of blunt-force trauma, with a Y-shaped depressed fracture running at the back of the bone. This injury was later identified as having been caused when the child's head was struck against the cot. Two areas of new bone formation were also identified on the skull as the result of previous blunt-force trauma and new bone formation on both femora indictaed where the child had been slapped. There was no evidence of rib fractures. The second case was of a child whose preliminary autopsy revealed nothing to dispute the parent's assertion that the child accidentally died while taking a bath. Osteological investigation revealed a hair-line fracture of the skull with an interrupted healing callus suggesting recurring trauma; dental fractures, new bone formation, fractures of the ulna and radius and an unusually high number of Harris lines ($n = 15$), suggestive of episodes of malnutrition and infection. The final case Walker (1997) reported was of a 7-year-old girl with multiple fractures to her nasal bones, tibia, radius, ulna, ribs and hands, and layered periosteal new bone on her cranium, mandible and long bones. In addition, the child had poor dental health consistent with studies that show a greater prevalence of dental disease in abused as opposed to non-abused children (Greene *et al.*, 1994), due to general neglect of their dental hygiene and a poor diet.

Table 8.2 *Skeletal lesions associated with child abuse*

Type of lesion	Location	Comments
Fractures	Ribs (51%) (Kleinman *et al.*, 1995)	First rib considered most severe and potentially pathognomic (Wilbur and Thompson, 1998; Starling, 2002)
	Long bones (44%) (Kleinman *et al.*, 1995)	Metaphyses affected in 89% cases (Kleinman *et al.*, 1995)
	Cranial blunt force (40%) (Wilbur and Thompson, 1998)	
	Hand and foot bones (4%) (Kleinman *et al.*, 1995; Resnick and Goergen, 2002)	
	Spine and clavicle (1%) (Kleinman *et al.*, 1995)	Fractures and dislocation of the spine may lead to kyphosis (Cullen, 1975)
	Dentition (Walker, 1997)	
	Acromion (Brogden, 1998)	
	Ilium (Wilbur and Thompson, 1998; Starling, 2002)	In severe trauma (Wilbur and Thompson, 1998; Starling, 2002)
New bone formation	Cranium (ecto- and endocranial) (Walker, 1997)	
	Mastoid process, superior manubrium, sternal end of clavicle (Saternus *et al.*, 2000)	Attachment of the sternocleidomastoid
	Proximal humerus (Saternus *et al.*, 2000)	Underlying the brachial plexus
	Proximal aspect of linea aspera (femur) (Saternus *et al.*, 2000)	Attachment of gluteus maximus
Radiographic features	'Bucket-handle' lesions at corners of metaphyses (Kleinman, 1987)	Areas of radiolucency with sclerotic margins
	Harris lines (Walker, 1997)	
Other	Dental caries, abscesses, calculus (Greene *et al.*, 1994)	

Table 8.2 and Fig. 8.7 summarise the type and distribution of skeletal lesions that might be expected in cases of child abuse. The examination of non-adult skeletal remains, particularly in forensic cases, must be done with knowledge of bone pathology and medical procedures that may have the appearances of child abuse. For example, birth trauma and instrument deliveries, skeletal dysplasias such as osteogenesis imperfecta, rickets, scurvy, cerebral palsy, congenital indifference to pain, infantile cortical hyperostosis and congenital syphilis may all mimic abuse (Brill and Winchester, 1987; Torwalt *et al.*, 2002). In many cases the presence of osteopenia and lack of rib fractures should help to distinguish pathological from non-accidental injuries.

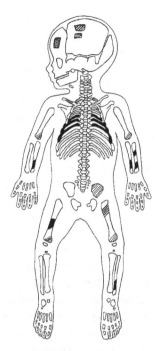

Figure 8.7 Distribution of lesions associated with child abuse in the non-adult skeleton. Black indicates most commonly affected areas; hatched lines indicate areas of less frequent involvement.

8.8 Child abuse in the archaeological record

Today, it is recognised that child abuse results from a complex combination of personal, social and cultural factors, such as generational violence, social stress and isolation, large family size, alcohol and drug abuse, poverty and mental illness. The child may also be at risk due to handicap, low birth weight and prematurity (Bethea, 1999). That child abuse is often seen in poor families may be due to the fact that wealthier families have the ability to hide the crime, due to less contact with social organisations. Whether the social conditions for child abuse existed in the past is debatable. However, urbanisation and industrialisation in the medieval and post-medieval periods resulting in the displacement of families into cities, and a greater social divide between the rich and poor, would have provided the circumstances in which abuse could occur, with tired and fractious parents returning home from work in dirty and often lethal factories. The fact that gin drinking in the lower classes was a serious social issue in the eighteenth century is well documented (Abel, 2001). Yet most post-medieval excavations are of the wealthy individuals who could afford to inter their children in family vaults, and who may have paid other people to care for their offspring on a daily basis. Up until the late eighteenth century, physical punishment of children was considered the norm, to the extent that 'heaven' was described to children as 'a place where children are never beaten' (Newman, 2004). In many societies children were kept away from their parents (e.g. Greco-Roman), and it is when they are living in closed contact with them that problems arise. Discovering what social situations existed in earlier undocumented periods is more of a challenge. Physical punishment was not frowned upon, but children may have been more seriously beaten by abusive individuals. That the Anglo-Saxons enjoyed drinking, sometimes to violent excess, is documented (Hagen, 1995), but violent acts towards children when under the influence are not. When urban and poorer cemeteries are excavated we need to examine the youngest children with a view to the discovery of possible physical abuse. But we need to know what to look for.

So why, if it is likely that child abuse took place in the past, and it seems from the literature that it did, do we not find any clear evidence of abuse in the skeletal record? There have been several explanations for this discrepancy. Waldron (2000) suggests that children who died from this form of abuse may have been buried in clandestine burials, rather than in the general cemetery. Walker (1997) suggests that the nature of physical abuse in the past was different from that reported in the forensic literature today, and that it was older children, working as apprentices and within factories, who may have been victims of abuse, rather than newborn children. In 1998, Blondiaux and colleagues reported a 2-year-old child from fourth-century Lisieux, Normandy, suspected of having suffered child abuse. Macroscopic, radiographic and histological examination revealed healed fractures on the parietal bones, endocranial lesions, perhaps suggestive of subdural haemorrhage in response to the trauma, ante-mortem loss of a mandibular incisor, and evidence of rickets that may indicate neglect.

It is possible that the practice of tightly swaddling children in seventeenth- and eighteenth-century Europe (Stone, 1977) may alter the pattern of lesions we would expect to see in abuse cases, making head trauma more likely and the characteristic postcranial fractures less common (Knight, 1986). De Mause (1974) refers to swaddled children being tossed between windows as a form of adult amusement, causing ribs to be crushed inwards as they were caught, although he later postulates that the passive nature of children in swaddling may have reduced the incidence of their abuse until the practice was abolished in the later eighteenth century.

The relatively small number of children suffering and dying from physical abuse today may have an impact on the numbers of children we would real- istically expect to see with such lesions in the archaeological record. Waldron (2000) suggests we would need to examine at least 2000 infant remains if we are to pick up just one child who may have suffered abuse in a given sample. With such small numbers of children surviving in the burial record, this is an impossible task. Diagnosis also relies on the complete preservation of the skeleton, including the ribs and cranial bones, and of course, the death of the infant. Many victims of abuse will survive and injuries will heal (unless there is severe damage to the epiphyses). The radiographic features used by pathologists to diagnose lesions on modern cases are also invisible to the bioarchaeologists, as hair-line fractures and metaphyseal 'chips' of bone are unlikely to be recognised in archaeological material. We do have the opportunity to record the more subtle changes invisible to radiographers, such as periosteal new bone formation indicating traumatic periosteal ripping and 'bone bruises'.

8.9   Summary

The study of skeletal trauma in the remains of children from past societies is limited, with only a handful of fractures being reported in the literature. The main reason for this deficit is the plastic nature of paediatric bone, which results in incomplete, greenstick fractures that heal rapidly without deformity, and the subtlety of other signatures of trauma such as premature epiphyseal fusion and periosteal lesions. The latter are often described as resulting from infection, and when widespread, assumed to be indicative of rickets and scurvy. For instance, periosteal cloaking, seen in victims of trauma including child abuse, is also a feature of congenital syphilis in the child. It is clear that, as today, children suffered from epiphyseal fractures, traumatic shortening of the limb and neurological injuries. The age at which these fractures occurred, their type and frequency can provide information on the age at which children began work, perhaps as apprentices, and were exposed to trauma. Evidence of child abuse is hindered by the often subtle and non-specific lesions that are left behind, and problems of differential diagnosis. However, it is the infants that cause the greatest challenges for trauma diagnosis. Infant remains are often very fragile and fragmentary, and are rarely recovered in their entirety. The careful analysis of infant remains in future should reveal evidence for birth injury and even child abuse, but crush fractures, linear fracture lines and cut marks are difficult to see in fragmentary cranial remains where the thin structure makes them difficult to reconstruct.

9 Future directions

In 1980, Buikstra and Cook stated that child studies were being hindered by poor preservation, lack of recovery and small sample sizes. Nearly 30 years on studies that focus on non-adult skeletal remains in contrasting populations, using a biocultural or bioarchaeological approach, are now more common. But we are still plagued by the same misconceptions and methodological limitations. There are still many challenges ahead before non-adult skeletons are regarded as being just as important as those of adults in providing information about the past. Their analysis provides opportunities for ever new and detailed questions to be addressed. It is an interesting time to be studying children, as more known-age and known-sex samples become available, and ever larger skeletal samples provide a large percentage of non-adult remains. The most notable collections include the known-age and known-sex samples from Christ Church Spitalfields and St Bride's Church in London and from Coimbra in Portugal. In the UK, the cemeteries of St Mary Spital, London and Barton-on-Humber, Bristol have yielded a large number of children and are now becoming more widely available for study. A few known-age children are also present in modern forensic samples, such as the Dart Collection in South Africa, and may provide a means for examining the growth and development of children from different ancestral groups. These endeavours are aided, or reflected, by the recent publication of high-quality texts on juvenile osteology which have increased the number of researchers familiar with their identification and complex skeletal anatomy (Scheuer and Black, 2000, 2004; Baker *et al.*, 2005).

There are many advantages to the study of non-adult skeletal remains. Age estimates are derived from specific markers, mainly of the dentition, which are linked to a genetic blueprint, unlike the more variable degenerative age indicators used for ageing adults. For this reason, non-adults can be aged more accurately than adults. The study of growth, rather than static adult stature, means that the age at which a group of children fall behind or recover their growth trajectory can be examined. Many indicators of stress on the skeleton and dentition develop during childhood, and bone lesions may remodel and disappear as the individual gets older. Therefore, the true prevalence of lesions such as cribra orbitalia and Harris lines can be more accurately assessed in younger individuals. Interpretations of these lesions are problematic and some authors have argued that they only develop in individuals who are strong enough to survive the stress episode. The precise age at which these lesions occur in childhood can help us answer questions about the nature of stress and survival in past populations. Nevertheless, children in an archaeological sample represent the non-survivors of any given population, and their pattern of growth or frequency of lesions might not be an accurate reflection of the living population of children who survived into adulthood. The early deaths of these children provide other challenges. Chronic diseases need time to develop on the skeleton, but children may die in the acute stages of the disease, before the skeleton has had time to respond. Their continued growth means that while conditions such as rickets will be more obvious when active, they disappear with equal rapidity, and incomplete (greenstick) fractures characteristic of non-adult trauma are particularly difficult to identify. Furthermore, we are still unable to provide a definitive sex from non-adult remains and are therefore missing vital information about the nature of infant mortality, the development of gender roles and changes in rates of growth and development.

That preservation of non-adult remains may be poor in some archaeological excavations is true. But this pattern is not universal and further, more detailed research needs to be carried out into the ages at which children's remains are less likely to survive in different burial conditions. All of these factors make the analysis of child skeletal remains one of the most challenging areas of research in both bioarchaeology and forensic anthropology. In this book, I have tried to explore the areas of study and matters of debate around the study of child skeletons; the rest of this chapter will outline some of the conclusions and future directions that can be drawn from this research.

9.1   Is absence of evidence evidence of absence?

Arguments that the remains of infants and children will never survive in the ground as well as those of adults are unfounded. In large cemetery excavations, non-adult skeletons can be found in sufficient numbers to allow in-depth research into their morbidity and mortality to be carried out. Evidence from the taphonomic literature suggests that, while children's bodies may decompose more quickly, their bones have the potential to survive well in the same conditions that allow for the good preservation of adult remains. In burial environments that make all human bones susceptible to decay, children's bones, being smaller and more difficult to identify, often suffer the most. Preliminary work on the density of non-adult bones shows that their cortical component changes through the child's life, with fetal bones being particularly dense, and cancellous bones, including the epiphyses, being especially susceptible to diagenesis. Modern research into bone porosity and permeability has only just begun to address the differences in these features in the growing skeleton.

Instead, the paucity of non-adults in a cemetery sample may be the result of other factors, including the skill of the excavator, curatorial bias and cultural exclusion. Funerary treatment of non-adults in the past does not follow any specific pattern. We have evidence for specific child cemeteries in Italy, Ireland and Wales, but have yet to uncover them in England, perhaps with the exception of the impromptu cluster at Hereford Cathedral. Where are the infants who appear to be missing from the early medieval cemeteries buried? Are all the Roman infants under houses? It is the reaction of each culture to the death of a child that effects whether, and how, they are buried. Children may have been buried in shallow graves, perhaps because of their smaller dimensions. Perhaps it was thought unnecessary to expend too much energy in digging a grave for these tiny remains. Unfortunately, few archaeological excavations consider grave depth or length in their analyses. Patterning of graves within communal cemeteries, where children are clustered in specific sections of the graveyard for practical or ritual reasons, will also limit their recovery if the entire cemetery is not excavated. When the marginalisation of infants does occur, it is of note and can tell us something about that society's attitude toward its children. Is there a specific age at which children begin to be incorporated into a cemetery? Is this because they are recognised as active members of the community, and do their gravegoods suggest they have taken on gender roles?

One section of the non-adult group that is often missing from our samples is adolescents. They have survived the most dangerous periods of their development and so rarely make it into the archaeological record in great numbers. A lack of these individuals on the brink of adulthood has frustrated research into rites of passage that may exist in certain societies. What was it like to be a 'teenager' in the past? Did such a category exist and what did it mean? Our understanding of the timing and extent of the puberty growth spurt in males and females is hindered by a lack of accurate sexing methods for this group. Although rare in archaeology, it is the older children that are more at risk of becoming forensic cases as runaways, suicide victims or armed combatants. Methods that now help us refine our ageing techniques in older children and young adults (e.g. annular ring fusion) are born out of the necessity to provide a biological and hopefully a positive identification in forensic cases.

9.2   Failure to adapt: children as non-survivors

The very fact that the children in the archaeological record died young informs us about the cultural and biological circumstances of a particular group. The ability of the adult population to keep the next generation, as vulnerable members of a community, alive and in good health is testament to their adaptability to their environment. The age at which children die is crucial to our understanding of what their lives were like, what diseases or trauma they were exposed to and why. Infants may die as the result of congenital defects, birthing practices, infanticide or exposure to infection in their new environment. Previously, clusters of infants in non-cemetery sites or hidden under the floors of houses were enough to convince archaeologists that infanticide was being practiced. Anthropological analysis using DNA, the neonatal line and a re-examination of our statistics has yet to produce unequivocal evidence for this practice. Children of weaning age are susceptible to disease, growth faltering and death if their weaning food is inadequate or contaminated. As they develop and begin to engage with their environment they may injure themselves, drown or become victims of abuse. Perhaps at 10 years old they begin to work, to be accepted as adults within a community, with occupational hazards exposing them to greater risks, but with wage-earning allowing them to become fully fledged and independent members of the society. Just as with adults, we need to be cautious when examining patterns of growth or frequency of lesions in children, in that they might not reflect the experience of the children who went on to survive into adulthood. But they were not all chronically sick children, many would have died as the result of an accident or acute infection and, up until their death, they would probably have experienced a childhood normal for that society.

9.3   Tom, Dick or Harriet? Sexing non-adults

Providing a sex estimate for non-adults continues to be the 'holy grail' in non-adult anthropology. Sex determinations based on the mandible are beginning to yield accuracy results as high as 82%, and those based on deciduous teeth dimensions can yield between 76% and 90%. But studies consistently show a male bias in the features analysed, and cut-off points in the age at which these features are no longer sexually dimorphic. Although sexual dimorphism has been identified in utero, there is still a disagreement about the validity of identifying morphological traits indicative of sex. While the application of a DNA analysis in determining the sex of non-adult skeletal material holds promise for the future, costs are still inhibitive, and we face the danger of reducing our samples to individual case studies based on DNA results, rather than estimating sex in a larger sample. This can be avoided in studies that test the reliability of certain morphological traits through DNA, allowing us to refine our methods. Just as with sex estimates, the identification of ancestry of a child is also influenced by morphological changes during growth and at puberty. In both biological and forensic anthropology, the development of isotopic fingerprinting to identify the country of origin of unknown individuals has become a significant tool. It may in time replace the traditional and problematic methods of ancestry assessment, particularly when tracing the geographical origins of non-adults who have yet to develop morphological ancestry traits.

9.4   The future

So what does the future hold for the skeletal analysis of non-adults? In bio-archaeology, recent advances in stable isotope analysis has begun a new era in the examination of weaning stress in past populations. It is now possible to iden-tify more precisely not only the age at which weaning occurred, but also what type of diet the child was weaned on to. Improvements in our ageing techniques, microscopic imaging of enamel defects and refinement of diagnostic criteria for rickets and scurvy have enabled us to provide a much more detailed picture of the interaction between childhood nutrition, infection and environmental change in past societies. If we are to continue to add to the new developments in the diagnosis of diseases in children, radiographs of the epiphyses should be included as features within them can provide diagnostic criteria for tuberculo-sis, osteomyelitis and syphilis. What is needed is an international database of collections that hold large numbers of non-adults (i.e. at least 100 individuals), of all ages and from different sites and periods to allow us to continue to add to the body of data on the child's experience.

We still have a long way to go in developing personal identification techniques for children in forensic anthropology. While progress has been made in refining our ageing techniques, especially for adolescents, we still struggle to develop data on ancestry or devise sexing techniques that can provide the 95% accuracy needed for the court of law. It seems unlikely that we will ever be able to build up a collection of willed bodies of children, and so we need to ensure that child forensic cases continue to be published in the academic literature in order for a body of data to be compiled.

The study of non-adult remains by bioarchaeologists began earlier than the theoretical concepts of childhood in social archaeology, and to a large extent they have been pursued independently of each other. Both have now reached a level of sophistication that should encourage communication and integration of the disciplines. The study of non-adult skeletal remains is finally gaining the recognition it deserves and it continues to be one of the most challenging and exciting areas of research in bioarchaeology.

References

Abel E. (2001) Gin Lane: did Hogarth know about fetal alcohol syndrome? *Alcohol and Alcoholism* **36**:131–134.

Acheson R. M. (1957) The Oxford method of assessing skeletal maturity. *Clinical Orthopaedics and Related Research* **10**:19–39.

Acheson R. M. (1959) Effects of starvation, septicaemia and chronic illness on the growth cartilage plate and metaphysis of the immature rat. *Journal of Anatomy* **93**:123–134.

Acheson R. M. and Macintyre M. N. (1958) The effects of acute infection and acute starvation on skeletal development. *British Journal of Experimental Pathology* **39**:37–45.

Acosta A., Goodman A., Backstrand J. and Dolphin A. (2003) Infants' enamel growth disruptions and the quantity and quality of mothers' perinatal diets in Solis, Mexico. *American Journal of Physical Anthropology* Suppl. **36**:55.

Acsádi G. and Nemeskéri J. (1970) *History of Human Lifespan and Mortality*. Budapest: Akademiai Kiado.

Adair L. (2004) Fetal adaptations to maternal nutritional status during pregnancy. *American Journal of Physical Anthropology* Suppl. **38**:50.

Adaline P., Piercecchi-Marti M. D., Bourlière-Najean B., *et al.* (2001) Postmortem assessment of fetal diaphyseal femoral length: validation of a radiographic methodology. *Journal of Forensic Sciences* **46**:215–219.

Adams B. and Byrd J. (2002) Interobserver variation of selected postcranial measurements. *Journal of Forensic Sciences* **47**:1193–1202.

Adams J. (1997) Radiology of rickets and osteomalacia. In D. Feldman, F. Glorieux and J. Pike (eds.) *Vitamin D*. New York: Academic Press, pp. 619–642.

Adams J. and Hamblen D. (1991) *Outline of Fractures*, 10th edn. London: Churchill-Livingstone.

Ağritmiş H., Yayci N., Colak B. and Aksoy E. (2004) Suicidal deaths in childhood and adolescence. *Forensic Science International* **142**:25–31.

Agarwal D., Agarwal K., Upadhyay S., *et al.* (1992) Physical and sexual growth pattern of affluent Indian children from 5 to 18 years of age. *Indian Pediatrics* **29**:1203–1282.

Akazawa T., Muhesen S., Dodo Y., Kondo O. and Mizoguichi Y. (1995) Neanderthal infant burial. *Nature* **377**:585–586.

Albanese J. (2002) The use of skeletal data for the study of secular change: methodological implications of combining data from different sources. *American Journal of Physical Anthropology* Suppl. **34**:36.

190   *References*

Albert A. (1998) The use of vertebral ring epiphyseal union for age estimation in two cases of unknown identity. *Forensic Science International* **97**:11–20.

Albert A. and Greene D. (1999) Bilateral asymmetry in skeletal growth and maturation as an indicator of environmental stress. *American Journal of Physical Anthropology* **110**:341–349.

Albert A. and Maples W. (1995) Stages of epiphyseal union for the thoracic and lumbar vertebral centra as a method of age determination for teenage and young adult skeletons. *Journal of Forensic Sciences* **40**:623–633.

Albert A. and McCallister K. (2004) Estimating age at death from thoracic and lumbar vertebral ring epiphyseal union data. *American Journal of Physical Anthropology* Suppl. **38**:51.

Albert A., Wescott D. and Sparks C. (2001) Bilateral asymmetry of epiphyseal union as an indicator of stress in the Arikara. *American Journal of Physical Anthropology* Suppl. **32**:31.

Alfonso M., Thompson J. and Standen V. (2003) Are Harris lines an indicator of stress? A comparison between Harris lines and enamel hypoplasia. *American Journal of Physical Anthropology* Suppl. **36**:58.

Allison M. J., Mendoza D. and Pezzia A. (1974) A radiographic approach to childhood illness in pre-Columbian inhabitants of Southern Peru. *American Journal of Physical Anthropology* **40**:409–416.

Al-Senan K., Al-Alaiyan S., Al-Abbad A. and LeQuesne G. (2001) Congenital rickets secondary to untreated maternal renal failure. *Journal of Perinatology* **21**:473–475.

Alt K., Alder C., Buitrago-Tellez C. and Lohrke B. (2002) Infant osteosarcoma. *International Journal of Osteoarchaeology* **12**:442–448.

Amirhakimi G. H. (1974) Growth from birth to two years for rich urban and poor rural Iranian children compared with Western norms. *Annals of Human Biology* **1**:427–442.

Andersen E., Andersen H., Hutchings B., *et al.* (1974) Hojde og vaegt hos danske skoleborn, 1971–1972. *Ugeskrift for Laeger* **136**:2796–2802.

Andersen E., Hutchings B., Jansen J. and Nyholm N. (1982) Hojde og vaegt hos danske born. *Ugeskrift for Laeger* **144**:1760–1765.

Anderson M. and Green W. (1948) Lengths of the femur and the tibia. *American Journal of Diseases in Children* **75**:279–290.

Anderson M., Messner M. B. and Green W. T. (1964) Distribution lengths of the normal femur and tibia in children from one to eighteen years of age. *Journal of Bone and Joint Surgery* **46**-A:1197–1202.

Anderson S. (1989) The human skeletal remains from The Hirsel, Coldstream, Berwickshire. Unpublished report. Durham, UK: Durham University and Historic Scotland.

Anderson T. (2001) The human remains. In M. Hicks and A. Hicks (eds.) *St Gregory's Priory, Northgate, Canterbury Excavations 1988–1991.* Canterbury, UK: Canterbury Archaeological Trust, pp. 338–370.

Anderson T. (2002) Metaphyseal fibrous defects in juveniles from medieval Norwich. *International Journal of Osteoarchaeology* **12**:144–148.

Anderson T. and Carter A. R. (1994) Periosteal reaction in a newborn child from Sheppey, Kent. *International Journal of Osteoarchaeology* **4**:47–48.

Anderson T. and Carter A. R. (1995) An unusual osteitic reaction in a young medieval child. *International Journal of Osteoarchaeology* 5:192–195.

Angel L. (1964) Osteoporosis; thalassemia? *American Journal of Physical Anthropology* 22:369–374.

Angel L. (1969) The basis of paleodemography. *American Journal of Physical Anthropology* 30:427–438.

Angel L. (1971) *The People of Lerna*. Washington, DC: Smithsonian Institution Press.

Ariès P. (1960) *Centuries of Childhood*. London: Jonathan Cape.

Armelagos G., Mielke J., Owen K., *et al.* (1972) Bone growth and development in prehistoric populations from Sudanese Nubia. *Journal of Human Evolution* 1:89–119.

Arneil G. (1973) Rickets in Glasgow today. *The Practitioner* 210:331–339.

Arriaza B., Allison M. and Gerszten E. (1988) Maternal mortality in Pre-Columbian Indians of Arica, Chile. *American Journal of Physical Anthropology* 77:35–41.

Asper H. (1916) Über die "Braune Retzius' Scheparallelstreifung" im Schmelz der menschlichen Zähne. *Schweizerische Viertelschrift für Zahneilkunde* 26:275.

Atkins P. J. (1992) White poison? The social consequences of milk consumption, 1850–1930. *Social History of Medicine* 15:207–226.

Attreed L. (1983) From pearl maiden to tower princes: towards a new history of medieval childhood. *Journal of Medieval History* 9:43–58.

Aufderheide A. C. and Rodriguez-Martin C. (1998) *The Cambridge Encyclopedia of Human Paleopathology*. Cambridge, UK: Cambridge University Press.

Aveling E. (1997) Chew, chew, that ancient chewing gum. *British Archaeology* 21:6–7.

Avishai G. and Smith P. (2002) The cremated infant remains from Carthage: skeletal and dental evidence for and against human sacrifice. *American Journal of Physical Anthropology* Suppl. 34:39.

Aykroyd R., Lucy D., Pollard A. and Solheim T. (1997) Technical note: regression analysis in adult age estimation. *American Journal of Physical Anthropology* 104:259–265.

Aykroyd R., Lucy D., Pollard A. and Roberts C. (1999) Nasty, brutish, but not necessarily short: a reconsideration of the statistical methods used to calculate age at death from adult human skeletal and dental age indicators. *American Antiquity* 64:55–70.

Aykroyd W. R. and Hossain M. A. (1967) Diet and state of nutrition of Pakistani infants in Bradford, Yorkshire. *British Medical Journal* 1:42–45.

Baart J. (1990) Ceramic consumption and supply in early modern Amsterdam; local production and long-distance trade. In P. Corfield and D. Keene (eds.) *Work in Towns 850–1850*. Leicester, UK: Leicester University Press, pp. 74–85.

Bagousse A.-L. and Blondiaux J. (2001) Hyperostoses corticales foetal et infantile à Lisieux (IVe s.): retour à Costebelle. *Centre Archaéologique du Var Revue* 2001:60–64.

Bailit H. and Hunt E. (1964) The sexing of children's skeletons from teeth alone and its genetic implications. *American Journal of Physical Anthropology* 22:171–174.

Baker B. and Armelagos G. J. (1988) Origin and antiquity of syphilis: a paleopathological diagnosis and interpretation. *Current Anthropology* 29:703–737.

Baker B., Dupras T. and Tocheris M. (2005) *The Osteology of Infants and Children*. College Station, TX: Texas A&M University Press.

Baker J. and Wright L. (1999) Introduction – childhood nutrition and health in prehistory. *American Journal of Physical Anthropology* Suppl. **28**:86–87.

Balthazard T. and Dervieux H. (1921) Etudes anthropologiques sur le foetus humain. *Annales Medecine Legale* **1**:37–42.

Barba W. P. and Freriks D. J. (1953) The familial occurrence of infantile cortical hyperostosis in utero. *Journal of Pediatrics* **42**:141–150.

Barker D. (1994) *Mothers, Babies and Disease in Later Life*. London: British Medical Journal Publishing Group.

Barlow T. (1883) On cases described as 'acute rickets' which are probably a combination of scurvy and rickets, the scurvy being an essential and the rickets a variable, element. *Medico-Chirurgical Transactions* **66**:159–219.

Barlow T. (1935) On cases described as 'acute rickets' which are probably a combination of scurvy and rickets. *Archives of Diseases in Childhood* **10**:223–252.

Barnes E. and Ortner D. (1997) Multifocal eosinophilic granuloma with a possible trepanation in a fourteenth century Greek young individual. *International Journal of Osteoarchaeology* **7**:542–547.

Baroncelli G., Bertelloni S., Ceccarelli C., Amato V. and Saggese G. (2000) Bone turnover in children with vitamin D deficiency rickets before and during treatment. *Acta Paediatrica* **89**:513–518.

Bass W., Evans D. and Jantz R. (1971) *The Leavenworth Site Cemetery: Archaeology and Physical Anthropology*. Lawrence, KS: University of Kansas Press.

Bassett E. (1982) Osteological analysis of Carrier Mills burials. In R. Jeffries and B. Butler (eds.) *The Carrier Mills Archaeological Project*. Carbondale, IL: Southern Illinois University, pp. 1029–1114.

Batey E. (2005) Subadult skeletal growth at Hierakonpolis, a working class cemetery of predynastic Egypt. *American Journal of Physical Anthropology* Suppl. **40**: 70.

Baxter J. (2005) *The Archaeology of Childhood: Children, Gender and Material Culture*. Walnut Creek, CA: Altamira Press.

Beard B. and Johnson C. (2000) Strontium isotope composition of skeletal material can determine the birth place and geographical mobility of humans and animals. *Journal of Forensic Sciences* **45**:1049–1061.

Beaton G. H. (1989) Small but healthy? Are we asking the right question? *Human Organization* **48**:30–39.

Beattie O. (1982) An assessment of x-ray energy spectroscopy and bone trace element analysis for the determination of sex from fragmentary human skeletons. *Canadian Journal of Anthropology* **2**:205–215.

Beausang E. (2000) Childbirth in prehistory: an introduction. *European Journal of Archaeology* **3**:69–87.

Behlmer G. K. (1979) Deadly motherhood: infanticide and medical opinion in mid-Victorian England. *Journal of the History of Medicine* **34**:403–427.

Beisel W. R. (1975) Synergistic effects of maternal malnutrition and infection on the infant. *American Journal of Diseases in Children* **129**:571–574.

Bell L., Skinner M. and Jones S. (1996) The speed of postmortem change to the human skeleton and its taphonomic significance. *Forensic Science International* **82**:129–140.

Bennike P., Lewis M., Schutkowski H. and Valentin F. (2005) Comparisons of child morbidity in two contrasting medieval cemeteries from Denmark. *American Journal of Physical Anthropology* **127**:734–746.

Berhrman R., Kliegman R. and Arvin A. (eds.) (1996) *Nelson Textbook of Pediatrics*, 15th edn. Philadelphia, PA: W. B. Saunders.

Bermudez de Castro J. and Rosas A. (2001) Pattern of dental development in Hominid XVIII from the Middle Pleistocene Atapuerca-Sima de los Huesos site (Spain). *American Journal of Physical Anthropology* **114**:352–330.

Bernard M. (2002) Tuberculosis in 20th century Britain: a preliminary study of the demographic profile of children admitted to Stannington sanatorium. *American Journal of Physical Anthropology* Suppl. **34**:44.

Bethea L. (1999) Primary prevention of child abuse. *American Family Physician* **59**:1577–1597.

Beunen G., Malina R., Van't Hof M., *et al.* (1988) *Adolescent Growth and Motor Performance: A Longitudinal Study of Belgian Boys.* Champaign, IL: Human Kinetics Books.

Billewicz W. and McGregor I. (1982) A birth to maturity longitudinal study of heights and weights in two West African (Gambian) villages, 1951–1975. *Annals of Human Biology* **9**:309–320.

Binns C. W. (1998) Infant-feeding and growth. In S. Ulijaszek, F. E. Johnston and M. A. Preece (eds.) *The Cambridge Encyclopedia of Human Growth and Development.* Cambridge, UK: Cambridge University Press, pp. 320–325.

Bird D. and Bird R. (2000) The ethnoarchaeology of juvenile foragers: shellfishing strategies among Meriam children. *Journal of Anthropological Archaeology* **19**:461–476.

Birkby W. and Gregg J. (1975) Otosclerotic stepedial footplate fixation in an eighteenth century burial. *American Journal of Physical Anthropology* **42**:81–84.

Bishop N. and Fewtrell M. (2003) Metabolic bone diseases of prematurity. In F. Glorieux, J. Pettifor and M. Jüppner (eds.) *Pediatric Bone: Biology and Diseases.* New York: Academic Press, pp. 567–581.

Black S. and Scheuer L. (1996) Occipitalization of the atlas with reference to its embryological development. *International Journal of Osteoarchaeology* **6**:189–194.

Black T. (1978) Sexual dimorphism in the tooth-crown diameters of the deciduous teeth. *American Journal of Physical Anthropology* **48**:77–82.

Blaha P. (1986) *Anthropometric Studies of the Czechoslovak Population: From 6 to 55 Years.* Prague: Czechoslovak Spartakiade.

Blakely R. L. (1989) Bone strontium in pregnant and lactating females from archaeological samples. *American Journal of Physical Anthropology* **80**:173–185.

Blakey M. L. and Armelagos G. J. (1985) Deciduous enamel defects in prehistoric Americans from Dickson Mounds: prenatal and postnatal stress. *American Journal of Physical Anthropology* **66**:371–380.

Blakey M. L., Leslie T. E. and Reidy J. P. (1994) Frequency and chronological distribution of dental enamel hypoplasia in enslaved African Americans: a test of the weaning hypothesis. *American Journal of Physical Anthropology* **95**:371–383.

Blanco R. A., Acheson R. M., Canosa C. and Salomon J. B. (1974) Height, weight and lines of arrested growth in young Guatemalan children. *American Journal of Physical Anthropology* **40**:39–48.

Blank E. (1975) Recurrent Caffey's cortical hyperostosis and persistent deformity. *Pediatrics* **55**:856–860.

Blom D., Buikstra J., Keng L., *et al.* (2005) Anemia and childhood mortality: latitudinal patterning along the coast of pre-Columbian Peru. *American Journal of Physical Anthropology* **127**:152–169.

Blondiaux G., Blondiaux J., Secousse F., *et al.* (1998) Rickets and child abuse: the case of a 4th century girl from Normandy. *Proceedings of the 12th European Meeting of the Paleopathology Association*, Prague, Czech Republic, p. 15.

Bocquet-Appel J. P. (2002) Paleoanthropological traces of a Neolithic demographic transition. *Current Anthropology* **43**:637–650.

Bocquet-Appel J. P. and Masset C. (1982) Farewell to paleodemography. *Journal of Human Evolution* **11**:321–333.

Bocquet-Appel J. P. and Masset C. (1996) Paleodemography: expectancy and false hope. *American Journal of Physical Anthropology* **99**:571–583.

Bogin B. (1988a) Rural-to-urban migration. In C. G. N. Mascie-Taylor and G. W. Lasker (eds.) *Biological Aspects of Human Migration*. Cambridge, UK: Cambridge University Press, pp. 90–129.

Bogin B. (1988b) *Patterns of Human Growth*. Cambridge, UK: Cambridge University Press.

Bogin B. (1991) Measurement of growth variability and environmental quality in Guatemalan children. *Annals of Human Biology* **18**:285–294.

Bogin B. (1997) Evolutionary hypotheses for human childhood. *Yearbook of Physical Anthropology* **40**:63–89.

Bogin B. (1998) Evolutionary and biological aspects of childhood. In C. Panter-Brick (ed.) *Biosocial Perspectives on Children*. Cambridge, UK: Cambridge University Press, pp. 10–44.

Bogin B. and Loucky J. (1997) Plasticity, political economy and physical growth status of Guatemala Maya children living in the United States. *American Journal of Physical Anthropology* **102**:17–32.

Bogin B., Wall M. and MacVean R. B. (1992) Longitudinal analysis of adolescent growth of Ladino and Mayan school children in Guatemala: effects of environment and sex. *American Journal of Physical Anthropology* **89**:447–457.

Bohnert M., Rost T., Faller-Marquardt M., Rodohl D. and Pollack S. (1997) Fractures of the base of the skull in charred bodies: postmortem heat injuries or signs of mechanical traumatisation? *Forensic Science International* **87**:55–62.

Bolaños M., Manrique M., Bolaños M. and Briones M. (2000) Approaches to chronological age assessment based on dental calcification. *Forensic Science International* **110**:97–106.

Bolaños M., Moussa H., Manrique M. and Bolaños M. (2003) Radiographic evaluation of third molar development in Spanish children and young people. *Forensic Science International* **133**: 212–219.

Bonnichsen R. (1973) Millie's Camp: an experiment in archaeology. *World Archaeology* **4**:277–291.

Boocock P., Roberts C. and Manchester K. (1995) Maxillary sinusitis in medieval Chichester, England. *American Journal of Physical Anthropology* **98**:483–496.

Borden S. (1974) Traumatic bowing of the forearm in children. *Journal of Bone and Joint Surgery* **56**-A:611–616.

Boric D. (2002) Apotropaism and the temporality of colourful Mesolithic–Neolithic seasons in the Danube Gorges. In A. Jones and G. MacGregor (eds.) *Colouring the Past: The Significance of Colour in Archaeological Research*. Oxford, UK: Berg, pp. 23–43.

Boucher B. (1955) Sex differences in the foetal sciatic notch. *Journal of Forensic Medicine* **2**:51–54.

Boucher B. (1957) Sex differences in the foetal pelvis. *American Journal of Physical Anthropology* **15**:581–600.

Bourgeois-Pichat J. (1951) La mesure de la mortalité infantile. I. Principes et méthodes. *Population* **6**:233–248.

Bouwman A. and Brown T. (2005) The limits of biomolecular palaeopathology: ancient DNA cannot be used to study venereal syphilis. *Journal of Archaeological Science* **32**:703–713.

Bowers E. (2005) Possible etiological significance of altered growth patterns in children with cleft of the lip and palate. *American Journal of Physical Anthropology* Suppl. **40**:76–77.

Boyde A. (1963) Estimating age at death of young human remains from incremental lines in dental enamel. *Proceedings of the 3rd International Meeting of Forensic Immunology, Medicine, Pathology and Toxicology*, London, 1962.

Boylston A., Wiggins R. and Roberts C. (1998) Human skeletal remains. In G. Drinkall and M. Foreman (eds.) *The Anglo-Saxon Cemetery at Castledyke South, Barton-on-Humber*. Sheffield, UK: Sheffield Excavation Reports, pp. 221–235.

Brahin J. and Fleming S. (1982) Children's health problems: some guidelines for their occurrence in ancient Egypt. *Museum of Applied Science Center for Archeology Journal* **2**:75–81.

Bramwell N. and Byard R. (1989) The bones in the Abbey: are they the murdered princes? *American Journal of Forensic Medicine and Pathology* **10**:83–87.

Briggs C. (1998) Anthropological assessment. In J. Clement and D. Ranson (eds.) *Craniofacial Identification in Forensic Medicine*. London: Arnold, pp. 49–61.

Brill P. W. and Winchester P. (1987) Differential diagnosis of child abuse. In P. Kleinmann (ed.) *Diagnostic Imaging of Child Abuse*. Baltimore, MD: Williams & Wilkins, pp. 221–241.

British Medical Association (1889) Survey on rickets. *British Medical Journal* **1**:114.

Britton H. A., Canby J. P. and Kohler C. M. (1960) Iron deficiency anemia producing evidence of marrow hyperplasia in the calvarium. *Pediatrics* **25**:621–628.

Brogdon B. (1998) Child abuse. In B. Brogdon (ed.) *Forensic Radiology*. New York: CRC Press, pp. 281–314.

Brooke O. G., Wood C. and Butters F. (1984) The body proportions for small-for-dates infants. *Early Human Development* **10**:85–94.

Brothwell D. (1971) Palaeodemography. In W. Brass (ed.) *Biological Aspects of Demography*. London: Taylor & Francis, pp. 111–128.

Brothwell D. (1981) *Digging Up Bones*. Ithaca, NY: Cornell University Press.

Brothwell D. (1986–7) The problem of the interpretation of child mortality in earlier populations. *Antropologia Portugesa* **4–5**:135 and 143.

Brothwell D. and Brown S. (1994) Pathology. In S. Lilley, G. Stroud, D. Brothwell and M. Williamson (eds.) *The Jewish Burial Ground at Jewbury.* York, UK: Council for British Archaeology, pp. 465–466.

Brothwell D., Powers R. and Wright S. (2000) Demography. In P. Rahtz, S. Hirst and S. Wright (eds.) *Cannington Cemetery: Excavations 1962–3 of Prehistoric, Roman, Post-Roman and Later Features at Cannington Park Quarry, Near Bridgewater, Somerset.* London: Society for the Promotion of Roman Studies, pp. 135–161.

Brown J. and Holden D. (2002) Iron acquisition by gram-positive bacterial pathogens. *Microbes and Infection* **4**:1149–1156.

Bruintjes T. (1990) The auditory ossicles in human skeletal remains from a leper cemetery in Chichester, England. *Journal of Archaeological Science* **17**:627–633.

Bruyère B. (1937) *Rapport sur les Fouilles de Deir el Médineh (1934–1935).* Cairo: Imprimerie de l'Institut Française d'Archéologie Orientale.

Buck T. and Strand Vidarsdottir U. (2004) A proposed method for the identification of race in sub-adult skeletons: a geometric morphometric analysis of mandibular morphology. *Journal of Forensic Sciences* **49**:1159–1164.

Buckberry J. (2000) *Missing Presumed Buried? Bone Diagenesis and the Under-representation of Anglo-Saxon Children.* Available online at http:/www.shef.ac.uk/~Cassem/5/buckberr.html/

Buckberry J. (2005) Where have all the children gone? The preservation of infant and children's remains in the archaeological record. Paper presented at the *Archaeology of Infancy and Childhood Conference*, 6–8 May 2005, University of Kent, UK.

Buckley H. (2000) Subadult health and disease in prehistoric Tonga, Polynesia. *American Journal of Physical Anthropology* **113**:481–505.

Buckley H. and Tayles N. (2003) Skeletal pathology in a prehistoric Pacific Island sample: issues in lesion recording, quantification, and interpretation. *American Journal of Physical Anthropology* **122**:303–324.

Buikstra J. (1976) *Hopewell in the Lower Illinois Valley.* Evanston, IL: Northwestern University Archeological Program.

Buikstra J. and Cook D. (1980) Palaeopathology: an American account. *Annual Review of Anthropology* **9**:433–470.

Buikstra J. E. and Konigsberg L. W. (1985) Paleodemography: critiques and controversies. *American Anthropologist* **87**:316–331.

Buikstra J. E. and Mielke J. H. (1985) Demography, diet, and health. In R. I. Gilbert and J. H. Mielke (eds.) *Analysis of Prehistoric Diets.* London: Academic Press, pp. 359–422.

Buikstra J. E. and Ubelaker D. (eds.) (1994) *Standards for Data Collection From Human Skeletal Remains.* Fayetteville, AR: Arkansas Archeological Survey.

Buikstra J., Konigsberg L. and Bullington J. (1986) Fertility and the development of agriculture in the prehistoric Midwest. *American Antiquity* **51**:191–204.

Burton J. and Wells M. (2002) The Alder Hey affair: implications for pathology practice. *Archives of Disease in Childhood* **86**:6–9.

Buzina R. (1976) Growth and development of three Yugoslav populations in different ecological settings. *American Journal of Clinical Nutrition* **29**:1051–1059.

Byers S. (1991) Technical note: calculation of age at formation of radiopaque transverse lines. *American Journal of Physical Anthropology* **85**:339–343.

Byers S. (2002) *Introduction to Forensic Anthropology: A Textbook*. Boston, MA: Allyn & Bacon.

Byers S. N. (1997) The relationship between stress markers and adult skeletal size. *American Journal of Physical Anthropology* Suppl. **24**:85–86.

Cadogan W. (1748) *Essay upon the Nursing and the Management of Children*. London.

Caffey J. (1939) Syphilis of the skeleton in early infancy. *American Journal of Roentgenology and Radium Therapy* **42**:637–655.

Caffey J. (1946) Multiple fractures in the long bones of infants suffering from chronic subdural hematoma. *American Journal of Roentgenology* **56**:163–173.

Caffey J. (1974) The Whiplash Shaken Infant Syndrome: manual shaking by the extremities with whiplash-induced intracranial and intraocular bleedings, linked with residual permanent brain damage and mental retardation. *Pediatrics* **54**:396–403.

Caffey J. (1978) *Pediatric X-Ray Diagnosis*. Chicago, IL: Tear Book Medical Publishers.

Caffey J. and Silverman W. A. (1945) Infantile cortical hyperostosis: preliminary report on a new syndrome. *American Journal of Roentgenology and Radium Therapy* **54**:1–16.

Cameron N. and Demerath E. (2002) Critical periods in human growth and their relationship to diseases of aging. *Yearbook of Physical Anthropology* **45**:159–184.

Canci A., Tarli S. M. B. and Repetto E. (1991) Osteomyelitis of probable haematogenous origin in a Bronze Age child from Toppo Daguzzo (Basilicata, Southern Italy). *International Journal of Osteoarchaeology* **1**:135–139.

Capellini E., Chiarelli B., Sineo L., *et al.* (2004) Biomolecular study of human remains from tomb 5859 in the Etruscan necropolis of Monterozzi, Tarquinia (Viterbo, Italy). *Journal of Archaeological Science* **31**:603–612.

Cappell D. F. and Anderson J. R. (1971) *Muir's Textbook of Pathology*. London: Edward Arnold.

Capucci E., Damiani S., Vernerando A., diRenzi A. and de Stefano G. (1982–3) Statura e peso in un campione della popolazione Italiana in accrescimento (4–15 anni). *Rivista di Antropologia* **62**:255–271.

Carli-Thiele P. (1995) Scurvy: investigations on the human skeleton using macroscopic and microscopic radiological methods. *Journal of Paleopathology* **7**:88.

Carli-Thiele P. (1996) *Spuren von Mangelerkrankungen en steinzeitlichen Kinderskeleten (Vestiges of Deficiency Diseases in Stone Age Child Skeletons)*. Gottingen, Germany: Verlag Erich Goltze.

Carty H. (1998) Children's sports injuries. *European Journal of Radiology* **26**:163–176.

Carvel D. (2002) Controversies concerning human tissue retention and implications for the forensic practitioner. *Journal of Clinical Forensic Medicine* **9**:53–60.

Castellana C. and Kósa F. (2001) Estimation of fetal age from dimensions of atlas and axis ossification centers. *Forensic Science International* **117**:31–43.

Cauwe N. (2001) Skeletons in motion, ancestors in action: early Mesolithic collective tombs in Southern Belgium. *Cambridge Archaeological Journal* **11**:147–163.

Chadwick Hawkes S. and Wells C. (1975) An Anglo-Saxon obstetric calamity from Kingsworthy, Hampshire. *Medical and Biological Illustration* **25**:47–51.

Chagula W. (1960) The age at eruption of third permanent molars in male East Africans. *American Journal of Physical Anthropology* **18**:77–82.

Chalke H. D. (1962) The impact of tuberculosis on history, literature and art. *Medical History* **6**:301–318.

Chamberlain A. (1997) In this dark cavern thy burial place. *British Archaeology* **26**:1–5.

Chen L. C. (1983) Interactions of diarrhea and malnutrition. In L. C. Chen and N. S. Scrimshaw (eds.) *Diarrhea and Malnutrition: Interactions, Mechanisms and Interventions*. New York: United Nations University/Plenum Press, pp. 3–19.

Chertkow S. (1980) Tooth mineralization as an indicator of the pubertal growth spurt. *American Journal of Orthodontics* **77**:79–91.

Chesterman J. (1979) Investigations of the human bones from Quanterness. In C. Renfrew (ed.) *Investigations in Orkney*. London: Society of Antiquaries, pp. 98–111.

Chesterman J. (1983) The human skeletal remains. In J. Hedges (ed.) *Isbister: A Chambered Tomb*, BAR (British Series) no. 115. Oxford: pp. 73–132.

Chick D. H. (1976) Study of rickets in Vienna 1919–1922. *Medical History* **20**:41–51.

Chierici R. and Vigi V. (1991) Dietary trace elements in early infancy. In R. Di Toro (ed.) *Infantile Nutrition: An Update*. Naples, Italy: Karger, pp. 66–85.

Clark J. and White W. (1998) Medieval bodies. In A. Werner (ed.) *London Bodies: The Changing Shape of Londoners from Prehistoric Times to the Present Day*. London: Museum of London, pp. 60–71.

Clarke S. (1977) Mortality trends in prehistoric populations. *Human Biology* **49**:181–186.

Clarke S. (1982) The association of early childhood enamel hypoplasias and radiopaque transverse lines in a culturally diverse prehistoric skeletal sample. *Human Biology* **54**:77–84.

Clement P. A., Bijiloos J., Kaufman L., *et al.* (1989) Incidence and aetiology of rhinosinusitis in children. *Acta Otorhinolaryngologica* **43**:523–543.

Cohen M. N. and Armelagos G. J. (eds.) (1984) *Paleopathology at the Origins of Agriculture*. New York: Academic Press.

Coleman E. (1976) Infanticide in the early Middle Ages. In S. M. Stuard (ed.) *Women in Medieval Society*. Philadelphia, PA: University of Pennsylvania Press, pp. 47–70.

Collins M., Nielsen-Marsh C., Hiller J., *et al.* (2002) The survival of organic matter in bone: a review. *Archaeometry* **44**:383–394.

Colson I. B., Richards M. B., Bailey J. F., Sykes B. C. and Hedges R. E. M. (1997) DNA analysis of seven human skeletons excavated from the Terp of Wijnaldum. *Journal of Archaeological Science* **24**:911–917.

Cook D. (1984) Subsistence and health in the Lower Illinois Valley: osteological evidence. In J. Cohen and G. Armelagos (eds.) *Paleopathology at the Origins of Agriculture*. Orlando, FL: Academic Press, pp. 237–271.

Cook D. (1994) Dental evidence for congenital syphilis (and its absence) before and after the conquest of the New World. In O. Dutour and G. Palfi (eds.) *The Origin of Syphilis in Europe: Before or After 1493?* Toulon, France: Université de Provence, pp. 169–175.

Cook D. (2005) Syphilis? Not quite: paleoepidemiology in an evolutionary context in the Midwest. In M. Powell and D. Cook (eds.) *The Myth of Syphilis: The Natural*

History of Treponematosis in North America. Gainesville, FL: University of Florida Press, pp. 177–199.

Cook D. and Buikstra J. E. (1979) Health and differential survival in prehistoric populations: prenatal dental defects. *American Journal of Physical Anthropology* **51**:649–664.

Cook D. and Powell M. (eds.) (2005) *The Myth of Syphilis: The Natural History of Treponematosis in North America.* Gainesville, FL: University of Florida Press.

Cool H. (2004) *The Roman Cemetery at Brougham, Cumbria: Excavations 1966–67.* London: Society for the Promotion of Roman Studies.

Cope Z. (1946) Fusion lines of bones. *Journal of Anatomy* **25**:280–281.

Corruccini R. S., Handler J. S. and Jacobi K. P. (1985) Chronological distribution of enamel hypoplasia and weaning in a Caribbean slave population. *Human Biology* **57**:699–711.

Coulon G. (1994) *L'Enfant en Gaule et Romaine.* Paris: Editions Errance.

Couper R., McPhee A. and Morris L. (2001) Indomethacin treatment of infantile cortical periostosis in twins. *Journal of Paediatrics and Child Health* **37**:305–308.

Coussens A., Anson T., Norris R. and Henneberg M. (2002) Sexual dimorphism in the robusticity of long bones of infants and young children. *Przeglad Antropologiczny* **65**:3–16.

Crawford S. (1991) When do Anglo-Saxon children count? *International Journal of Theoretical Archaeology* **2**:17–24.

Crawford S. (1993) Children, death and the afterlife in Anglo-Saxon England. *Anglo-Saxon Studies in Archaeology and History* **6**:83–91.

Crawford S. (1999) *Childhood in Anglo-Saxon England.* Stroud, UK: Alan Sutton Publishing.

Crist T. A., Washburn A., Park H., Hood I. and Hickey M. (1997) Cranial bone displacement as a taphonomic process in potential child abuse cases. In W. Haglund and M. Sorg (eds.) *Forensic Taphonomy: The Postmortem Fate of Human Remains.* New York: CRC Press, pp. 319–336.

Cronjé G. (1984) Tuberculosis and mortality decline in England and Wales, 1851–1910. In R. Woods and J. Woodward (eds.) *Urban Disease and Mortality in Nineteenth-Century England.* London: Batsford, pp. 79–101.

Crooks D. L. (1999) Child growth and nutritional status in a high-poverty community in Eastern Kentucky. *American Journal of Physical Anthropology* **109**:129–142.

Cullen J. C. (1975) Spinal lesions of battered babies. *Journal of Bone and Joint Surgery* **57**-B:364–366.

Cumming W. (1979) Neonatal skeletal fractures: birth trauma or child abuse? *Journal de l'Association Canadienne des Radiologistes* **30**:30–33.

Cunha E., Fily M.-L., Clisson I., *et al.* (2000) Children at the convent: comparing historical data, morphology and DNA extracted from ancient tissues for sex diagnosis at Santa Clara-a-Velha (Coimbra, Portugal). *Journal of Archaeological Science* **27**:949–952.

Cunha E., Rozzi F., Bermúdez de Castro J., *et al.* (2004) Enamel hypoplasias and physiological stress in the Sima de los Huesos Middle Pleistocene hominins. *American Journal of Physical Anthropology* **125**:220–231.

Cunningham H. (1995) *Children and Childhood in Western Society since 1500*. Harlow, UK: Longman.

Currey J. and Butler G. (1975) The mechanical properties of bone tissue in children. *Journal of Bone and Joint Surgery* **57**-A:810–814.

Curtin A. (2005) Prehistoric treponematosis in the Pacific Northwest. In M. Powell and D. Cook (eds.) *The Myth of Syphilis: The Natural History of Treponematosis in North America*. Gainesville, FL: University of Florida Press, pp. 306–330.

Curwen C. (1934) Excavations in Whitehawk Neolithic Camp, Brighton, 1932–3. *Antiquaries' Journal* **14**:123–126.

Dahlberg A. and Menegaz-Bock R. (1958) Emergence of the permanent teeth in Pima Indian children. *Journal of Dental Research* **37**:1123–1140.

Dalby G., Manchester K. and Roberts C. A. (1993) Otosclerosis and stapedial footplate fixation in archaeological material. *International Journal of Osteoarchaeology* **3**:207–212.

Dallman P. R., Simes M. A. and Stekel A. (1980) Iron deficiency in infancy and childhood. *American Journal of Clinical Nutrition* **33**:86–118.

Daly R., Saxon L., Turner C., Robling A. and Bass S. (2004) The relationship between muscle size and bone geometry during growth and in response to exercise. *Bone* **34**:281–287.

Danforth M. and Jacobi K. (2003) Patterns of correlation among morphological traits in the deciduous and permanent dentitions of juveniles. *American Journal of Physical Anthropology* Suppl. **36**:83.

Daniell C. (1997) *Death and Burial in Medieval England 1066–1550*. London: Routledge.

Davies C. T. M., Mbelwa D. and Doré C. (1974) Physical growth and development of urban and rural east African children, aged 7–16 years. *Annals of Human Biology* **1**:257–268.

Davies D. and Parsons F. (1927) The age order of the appearance and union of the normal epiphyses as seen by X-rays. *Journal of Anatomy* **62**:58–71.

Davis P. and Hägg U. (1994) The accuracy and precision of the "Demirjian System" when used for age determination in Chinese children. *Swedish Dental Journal* **18**:113–116.

Davidson J., Rose J., Gurmann M., *et al.* (2002) The quality of African-American life in the old southwest near the turn of the twentieth century. In R. H. Steckel and J. C. Rose (eds.) *The Backbone of History: Health and Nutrition in the Western Hemisphere*. Cambridge, UK: Cambridge University Press, pp. 226–280.

Dawes J. D. and Magilton J. R. (1980) *The Cemetery of St. Helen-on-the-Walls, Aldwark*. London: Council for British Archaeology.

Dawson L., Levy T. and Smith P. (2003) Evidence of interpersonal violence at the Chalcolithic village of Shiqmim (Israel). *International Journal of Osteoarchaeology* **13**:115–119.

De la Rúa C., Izagirre N. and Manzano C. (1995) Environmental stress in a medieval population of the Basque country. *Homo* **45**:268–289.

De Mause L. (1974) *The History of Childhood*. New York: Psychohistory Press.

De Vito C. and Saunders S. (1990) A discriminant function analysis of deciduous teeth to determine sex. *Journal of Forensic Sciences* **35**:845–858.

Deedrick D. (2000) Hair, fibers, crime and evidence. I Hair evidence. *Forensic Science Communications* **2**(3), available online at http://www.fbi.gov/hq/lab/fsc/current/index.htm/

Delgado-Rodríguez M., Pérez-Iglesias R., Gómez-Olmedo M., Bueno-Cavanillas A. and Gálvez-Vargas R. (1998) Risk factors for low birth weight: results from a case-control study in Southern Spain. *American Journal of Physical Anthropology* **105**:419–424.

DeLucia M. and Carpenter T. (2002) Rickets in the sunshine? *Nutrition* **18**:97–98.

Demirjian A. (1990) Dentition. In F. Falkner and J. M. Tanner (eds.) *Human Growth: A Comprehensive Treatise*. New York: Plenum Press, pp. 269–297.

Demirjian A. and Goldstein H. (1976) New systems for dental maturity based on seven and four teeth. *Annals of Human Biology* **3**:411–421.

Demirjian A. and Levesque G. Y. (1980) Sexual differences in dental development and prediction of emergence. *Journal of Dental Research* **59**:1110–1122.

Demirjian A., Goldstein H. and Tanner J. M. (1973) A new system of dental age assessment. *Human Biology* **45**:211–227.

Dennison J. (1979) Citrate estimation as a means of determining the sex of human skeletal material. *Archaeology and Physical Anthropology in Oceania* **14**:136–143.

Derevenski J. S. (ed.) (1994) Perspectives on Children and Childhood. *Archaeological Review from Cambridge* **13**(2).

Derevenski J. S. (1997) Engendering children, engendering archaeology. In J. Moore and E. Scott (eds.) *Invisible People and Processes*. Leicester, UK: Leicester University Press, pp. 192–202.

Derevenski J. S. (2000) Material culture shock: confronting expectations in the material culture of children. In J. S. Derevenski (ed.) *Children and Material Culture*. London: Routledge, pp. 1–16.

Dettwyler K. A. (1992) Nutritional status of adults in rural Mali. *American Journal of Physical Anthropology* **88**:309–321.

Dettwyler K. A. (1995) A time to wean: the hominid blueprint for the natural age of weaning in modern human populations. In P. Stuart-Macadam and K. Dettwyler (eds.) *Breastfeeding: Biocultural Perspectives*. New York: Aldine de Gruyter, pp. 39–73.

Dettwyler K. A. and Fishman C. (1992) Infant feeding practices and growth. *Annual Review of Anthropology* **21**:171–204.

Deutsch D., Tam O. and Stack M. (1985) Postnatal changes in size, morphology and weight of developing postnatal deciduous anterior teeth. *Growth* **49**:202–217.

Di Maio V. J. M. (1999) *Gunshot Wounds: Practical Aspects of Firearms, Ballistics, and Forensic Techniques*. New York: CRC Press.

Dirkmaat D. (2002) Recovery and interpretation of the fatal fire victim: the role of forensic anthropology. In W. Haglund and M. Sorg (eds.) *Advances in Forensic Taphonomy: Method, Theory and Archaeological Perspectives*. New York: CRC Press, pp. 451–472.

Divale W. (1972) Systematic population control in the Middle and Upper Palaeolithic: inferences based on contemporary hunter–gatherers. *World Archaeology* **4**:222–243.

Dolphin A. and Goodman A. (2002) The influence of maternal diet on pregnancy-forming enamel zinc concentrations of children from the Solis Valley, Mexico. *American Journal of Physical Anthropology* Suppl. **34**:64.

Dreizen S., Currie C., Gilley E. J. and Spies T. D. (1959) Observations on the association between nutritive failure, skeletal maturation rate and radiopaque transverse lines in the radius in children. *American Journal of Roentgeneology* **76**:482–487.

Dreizen S., Spirakis C. N. and Stone R. E. (1964) The influence of age and nutritional status on 'bone scar' formation in the distal end of the growing radius. *American Journal of Physical Anthropology* **22**:295–306.

Drinkhall G. and Foreman M. (1998) *The Anglo-Saxon Cemetery at Castledyke South, Barton-on-Humber.* Sheffield, UK: Humberside Archaeological Partnership.

Dufka C. (1999) Children as killers. In R. Gutman and D. Rieff (eds.) *Crimes of War: What the Public Should Know.* New York: W. W. Norton, pp. 78–79.

Dufour D. and Sauther M. (2002) Comparative and evolutionary dimensions of the energetics of human pregnancy and lactation. *American Journal of Human Biology* **14**:584–602.

Duggan A. and Wells C. (1964) Four cases of archaic disease of the orbit. *Eye, Ear, Nose and Throat Digest* **26**:63–68.

Duhig C. (1998) The human skeletal material. In T. Malim and J. Hines (eds.) *The Anglo-Saxon Cemetery at Edix Hill (Barrington A), Cambridgeshire.* York, UK: Council for British Archaeology, pp. 154–199.

Dunlop O. (1912) *English Apprenticeship and Child Labour.* London: T. Fisher Unwin.

Dunn A. W. and Aponte G. E. (1962) Congenital bowing of the tibia and femur. *Journal of Bone and Joint Surgery* **44**-A:737–740.

Dupras T., Schwartz H. and Fairgrieve S. I. (2001) Infant feeding and weaning practices in Roman Egypt. *American Journal of Physical Anthropology* **115**:204–212.

Duray S. M. (1996) Dental indicators of stress and reduced age at death in prehistoric Native Americans. *American Journal of Physical Anthropology* **99**:275–286.

Dvonch V. and Bunch W. (1983) Pattern of closure of the proximal femoral and tibial epiphyses in man. *Journal of Pediatric Orthopedics* **3**:498–501.

Dye T. J., Lucy D. and Pollard A. M. (1995) The occurrence and implications of post-mortem 'pink teeth' in forensic and archaeological cases. *International Journal of Osteoarchaeology* **5**:339–348.

Dyhouse C. (1978) Working-class mothers and infant mortality rates in England, 1895–1914. *Journal of Social History* **12**:248–267.

EAAF (1995) *Argentine Forensic Anthropology Team Report: Guatemala.* Available online at http://www.eaaf.org.ar/guatemala_eng-1995.htm/

East A. (2003) Normal periosteal bone growth and skeletal pathology in documented fetuses, University of New Mexico, Maxwell Museum documented collection and University of Tennessee documented collection. *American Journal of Physical Anthropology* Suppl. **32**:115.

East A. and Buikstra J. (2001) Is the fetus from Elizabeth Mound 3, Lower Illinois River Valley an achrondroplastic dwarf? *American Journal of Physical Anthropology* Suppl. **32**:61.

Eccles A. (1977) Obstetrics in the 17th and 18th centuries and its implications for maternal and infant mortality. *Bulletin for the Society of the Social History of Medicine* **20**:8–11.

Eccles A. (1982) *Obstetrics and Gynaecology in Tudor and Stuart England.* London: Croom Helm.

Edgar H. and Lease L. (2005) Comparison of deciduous and permanent dental morphology in a European American sample. *American Journal of Physical Anthropology* Suppl. **40**:98–99.

Ehrenberg M. (1989) *Women in Prehistory.* London: British Museum Publications.

Eiben O. and Panto E. (1986) The Hungarian National Growth Standards. *Anthropologai Kozlemenyek* **30**:1–40.

Eli I., Sarnet H. and Talmi E. (1989) Effect of the birth process on the neonatal line in primary tooth enamel. *Pediatric Dentistry* **11**:220–223.

El-Najjar M. Y. (1977a) Maize, malaria and the anemias in the pre-Columbian New World. *Yearbook of Physical Anthropology* **20**:329–337.

El-Najjar M. Y. (1977b) Porotic hyperostosis in North America: a theory. In E. Cockburn (ed.) *Porotic Hyperostosis: An Enquiry.* Detroit, MI: Paleopathology Association, pp. 9–10.

El-Najjar M. Y., Ryan D. J., Turner C. G. II and Lozoff B. (1979) The etiology of porotic hyperostosis among the prehistoric and historic Anasazi Indians of southwestern United States. *American Journal of Physical Anthropology* **44**:477–488.

Evans A. (2002) An evaluation of the effect of sport on the immature skeleton and its potential for individualisation. M.Sc. dissertation, Bournemouth University, Bournemouth, UK.

Eveleth P. B. and Tanner J. M. (1990) *Worldwide Variation in Human Growth.* Cambridge, UK: Cambridge University Press.

Faerman M. (1999) Ancient DNA diagnosis of bone pathology in infancy and early childhood. *American Journal of Physical Anthropology* Suppl. **28**:125.

Faerman M., Filon D., Kahila G., *et al.* (1995) Sex identification of archaeological human remains based on amplification of the X and Y amelogenin alleles. *Gene* **167**:327–332.

Faerman M., Jankaukas R., Gorski A., Bercovier H. and Greenblatt C. (1999) Detecting *Mycobacterium tuberculosis* in medieval skeletal remains from Lithuania. In G. Pálfi, O. Dutour, J. Deák and I. Hutás (eds.) *Tuberculosis: Past and Present.* Budapest: Golden Book Publishers and Tuberculosis Foundation, pp. 371–376.

Faerman M., Kahila Bar-Gal G., Filon D., *et al.* (1998) Determining the sex of infanticide victims from the late Roman era through ancient DNA analysis. *Journal of Archaeological Science* **25**:861–865.

Faerman M., Kahila G., Smith P., *et al.* (1997) DNA analysis reveals the sex of infanticide victims. *Nature* **385**:212–213.

Fairgrieve S. I. and Molto J. (2000) Cribra orbitalia in two temporally disjunct population samples from the Dakhleh Oasis, Egypt. *American Journal of Physical Anthropology* **111**:319–331.

Falkner F., Holzgreve W. and Schloo R. (2003) *Prenatal Influences on Postnatal Growth: Overview and Pointers for Needed Research.* Available online at http://www.unu.edu/unupress/food2.htm/

Fanning E. and Moorrees C. (1969) A comparison of permanent mandibular molar formation in Australian Aborigines and Caucasoids. *Archives of Oral Biology* **14**:999–1006.

Farwell D. E. and Molleson T. I. (1993) *Excavations at Poundbury 1966–80*, vol. 2, *The Cemeteries*. Dorchester, UK: Dorset Natural History and Archaeological Society.

Fazekas I. G. and Kósa F. (1978) *Forensic Fetal Osteology*. Budapest: Academic Press.

Feldesman M. R. (1992) Femur/stature ratio and estimates of stature in children. *American Journal of Physical Anthropology* **87**:459–477.

Ferencz M. and Józsa L. (1992) Congenital syphilis on a medieval skeleton. *Anthropologie* **30**:95–98.

Fildes V. (1986a) *Breasts, Bottles and Babies: A History of Infant Feeding*. Edinburgh, UK: Edinburgh University Press.

Fildes V. (1986b) 'The English Disease': infantile rickets and scurvy in pre-industrial England. In J. Cule and T. Turner (eds.) *Child Care through the Centuries*. London: British Society for the History of Medicine, pp. 121–134.

Fildes V. (1988a) *Wet Nursing: A History from Antiquity to the Present*. New York: Basil Blackwell.

Fildes V. (1988b) The English wet-nurse and her role in infant care 1538–1800. *Medical History* **32**:142–173.

Fildes V. (1992) Breast-feeding in London, 1905–19. *Journal of Biosocial Sciences* **24**:53–70.

Fildes V. (1995) The culture and biology of breastfeeding: an historical overview of Western Europe. In P. Stuart-Macadam and K. A. Dettwyler (eds.) *Breastfeeding: Biocultural Perspectives*. New York: Aldine de Gruyter, pp. 76–101.

Fildes V. (1998) Infant feeding practices and infant mortality in England, 1900–1919. *Continuity and Change* **13**:251–280.

Finlay N. (1997) Kid napping: the missing children in lithic analysis. In J. Moore and E. Scott (eds.) *Invisible People and Processes*. Leicester, UK: Leicester University Press, pp. 203–212.

Finlay N. (2000) Outside of life: traditions of infant burial in Ireland from cíllín to cist. *World Archaeology* **31**:407–422.

Finsterbush A. and Husseini N. (1979) Infantile cortical hyperostosis with unusual clinical manifestations. *Clinical Orthopaedics and Related Research* **144**:276–279.

Finucane R. (1997) *The Rescue of the Innocents: Endangered Children in Medieval Miracles*. London: Macmillan.

Fitzgerald C. M. and Saunders S. (2005) Test of histological methods of determining chronology of accentuated striae in deciduous teeth. *American Journal of Physical Anthropology* **127**:277–290.

Flecker H. (1932) Roentgenographic observations of the times of appearance of epiphyses and their fusion with diaphyses. *Journal of Anatomy* **67**:118–164.

Flecker H. (1942) Time of appearance and fusion of ossification centres as observed by roentgenographic methods. *American Journal of Roentgenology and Radium Therapy* **47**:97–159.

Floud R. and Wachter K. (1982) Poverty and physical stature: evidence on the standard of living of London boys, 1770–1870. *Social Science History* **6**:422–452.

Fogel M., Tuross N. and Owsley D. (1989) Nitrogen isotope tracers of human lactation in modern and archaeological populations. *Carnegie Institute of Washington Yearbook* **88**:111–117.

Follis R. H. and Park E. A. (1952) Some observations on bone growth, with particular respect to zones of increased density in the metaphysis. *American Journal of Roentgenology, Radium Therapy and Nuclear Medicine* **68**:709–724.

Fomon S. (1967) *Infant Nutrition*. Philadelphia, PA: W. B. Saunders.

Foote K. and Marriott L. (2003) Weaning of infants. *Archives of Diseases in Childhood* **88**:488–492.

Forbes T. R. (1986) Deadly parents: child homicide in eighteenth and nineteenth century England. *Journal of History of Medicine and Allied Sciences* **41**:175–199.

Formicola V. and Buzhilova A. (2004) Double child burial from Sunghir (Russia): pathology and inferences for Upper Palaeolithic funerary practices. *American Journal of Physical Anthropology* **124**:189–198.

Foti B., Lalys L., Adalian P., *et al.* (2003) New forensic approach to age determination in children based on tooth eruption. *Forensic Science International* **132**:1–8.

Fournier A. (1884) Syphilitic teeth. *Dental Cosmos* **26**:12–25, 141–155.

Fox P., Elston M. and Waterlow J. (1981) Pre-school child survey. In Department of Health and Social Security (ed.) *Subcommittee on Nutritional Surveillance: Second Report*. London: Her Majesty's Stationery Office.

Fraser D. R. (1995) Vitamin D. *Lancet* **345**:104–105.

Frisancho A. R. and Baker P. T. (1970) Altitude and growth: a study of the pattern of physical growth of a high altitude Peruvian Quechua population. *American Journal of Physical Anthropology* **32**:279–292.

Frisancho A. R., Garn S. M. and Ascoli W. (1970a) Childhood retardation resulting in reduction of adult body size due to lesser adolescent skeletal delay. *American Journal of Physical Anthropology* **33**:325–336.

Frisancho A. R., Garn S. M. and Ascoli W. (1970b) Unequal influence of low dietary intake on skeletal maturation during childhood and adolescence. *American Journal of Clinical Nutrition* **23**:1220–1227.

Frisancho A. R., Klayman J. E. and Matos J. (1977) Influence of maternal nutritional status on prenatal growth in a Peruvian urban population. *American Journal of Physical Anthropology* **46**:265–274.

Frisancho A. R., Matos J., Leonard W. R. and Yaroch L. A. (1985) Developmental and nutritional determinants of pregnancy outcome among teenagers. *American Journal of Physical Anthropology* **66**:247–261.

Frisancho A. R., Sanchez J., Pallardel D. and Yanez L. (1973) Adaptive significance of small body size under poor socio-economic conditions in southern Peru. *American Journal of Physical Anthropology* **39**:255–262.

Frost H. M. (1964) Dynamics of bone remodelling. In H. M. Frost (ed.) *Bone Biodynamics*. London: J. & A. Churchill, pp. 315–333.

Fuller B., Richards M. and Mays S. (2003) Stable carbon and nitrogen isotope variations in tooth dentine serial sections from Wharram Percy. *Journal of Archaeological Science* **30**:1673–1684.

Fuller B., Fuller J., Dage N., *et al.* (2004) Nitrogen balance and δ^{15}N: why you're not what you eat during pregnancy. *Rapid Communications in Mass Spectroscopy* **18**:2889–2896.

Fyllingen H. (2003) Society and violence in the Early Bronze Age: an analysis of human skeletons from Nord-Trondelag, Norway. *Norwegian Archaeological Review* **36**:27–43.

Garn S. and Baby R. S. (1969) Bilateral symmetry in finer lines of increased density. *American Journal of Physical Anthropology* **31**:89–92.

Garn S. and Clark D. C. (1976) Problems in the nutritional assessment of Black individuals. *American Journal of Public Health* **66**:262–267.

Garn S., Lewis A., Swindler D. and Kerewsky R. (1967) Genetic control of sexual dimorphism in tooth size. *Journal of Dental Research* **46**:963–972.

Garn S., Sandusky S., Nagy J. and Trowbridge F. (1973) Negro–Caucasoid differences in permanent tooth emergence at a constant income level. *Archives of Oral Biology* **18**:609–615.

Garn S., Osborne R. and McCabe K. (1979) The effect of prenatal factors on crown dimensions. *American Journal of Physical Anthropology* **51**:665–678.

Geddes J. and Whitwell H. (2004) Inflicted head injury in infants. *Forensic Science International* **146**:83–88.

Gelman A., Carlin J., Stern H. and Rubin D. (1995) *Bayesian Data Analysis*. London: Chapman & Hall.

Genc M. and Ledger W. (2005) Syphilis in pregnancy. *Sexually Transmitted Infections* **76**:73–79.

Gilbert R. I. and Mielke J. M. (eds.) (1985) *The Analysis of Prehistoric Diets*. Orlando, FL: Academic Press.

Gilchrist R. (2004) Archaeology and the life-course: a time and age for gender. In L. Meskell and R. Preucel (eds.) *A Companion to Social Archaeology*. Oxford, UK: Blackwell Publishing, pp. 142–160.

Gilchrist R. and Sloane B. (2005) *Requiem: The Monastic Cemeteries in Britain*. London: Museum of London.

Gillett R. M. (1997) Dental emergence among urban Zambian school children: an assessment of the accuracy of three methods in assigning ages. *American Journal of Physical Anthropology* **102**:447–454.

Gindhart S. P. (1969) The frequency of appearance of transverse lines in the tibia in relation to childhood illnesses. *American Journal of Physical Anthropology* **31**:17–22.

Gindhart P. S. (1973) Growth standards for the tibia and radius in children aged one month through eighteen years. *American Journal of Physical Anthropology* **39**:41–48.

Gladykowska-Rzeczycka J. J. and Krenz M. (1995) Extensive change within a subadult skeleton from a medieval cemetery of Slaboszewo, Mogilno district, Poland. *Journal of Paleopathology* **7**:177–184.

Gleiser I. and Hunt E. (1955) The permanent mandibular first molar: its calcification, eruption and decay. *American Journal of Physical Anthropology* **13**:253–281.

Glencross B. and Stuart-Macadam P. (2000) Childhood trauma in the archaeological record. *International Journal of Osteoarchaeology* **10**:198–209.

Glencross B. and Stuart-Macadam P. (2001) Radiographic clues to fractures of distal humerus in archaeological remains. *International Journal of Osteoarchaeology* 11:298–310.

Glen-Haduch E., Szostek K. and Glab H. (1997) Cribra orbitalia and trace element content in human teeth from Neolithic and Early Bronze Age graves in southern Poland. *American Journal of Physical Anthropology* 103:201–207.

Gleser K. (1949) Double contour, cupping and spurring in roentgenograms of long bones in infants. *American Journal of Roentgenology and Radium Therapy* 61:482–492.

Goldberg P. J. P. (1986) Female labour, service and marriage in the late medieval urban north. *Northern History* 12:18–38.

Goode H., Waldron T. and Rogers J. (1993) Bone growth in juveniles: a methodological note. *International Journal of Osteoarchaeology* 3:321–323.

Goodman A. H. (1993) On the interpretation of health from skeletal remains. *Current Anthropology* 34:281–288.

Goodman A. H. and Armelagos G. J. (1985) Factors affecting the distribution of enamel hypoplasias within the human permanent dentition. *American Journal of Physical Anthropology* 68:479–493.

Goodman A. H. and Armelagos G. J. (1988) Childhood stress and decreased longevity in a prehistoric population. *American Anthropology* 90:936–944.

Goodman A. H. and Armelagos G. J. (1989) Infant and childhood morbidity and mortality risks in archaeological populations. *World Archaeology* 21:225–243.

Goodman A. H. and Rose J. C. (1990) Assessment of systemic physiological perturbations from dental enamel hypoplasias and associated histological structures. *Yearbook of Physical Anthropology* 33:59–110.

Goodman A. H. and Rose J. C. (1991) Dental enamel hypoplasias as indicators of nutritional status. In M. A. Kelley and C. Larsen (eds.) *Advances in Dental Anthropology*. New York: Wiley-Liss, pp. 279–293.

Goodman A. H., Allen L. H., Hernandez G. P., *et al.* (1987) Prevalence and age at development of enamel hypoplasias in Mexican children. *American Journal of Physical Anthropology* 72:7–19.

Goodman A. H., Brooke-Thomas R., Swedlund A. C. and Armelagos G. J. (1988) Biocultural perspectives on stress in prehistoric, historical and contemporary population research. *Yearbook of Physical Anthropology* 31:169–202.

Gordon C. C. and Buikstra J. (1981) Soil pH, bone preservation and sampling bias at mortuary sites. *American Antiquity* 48:566–571.

Gordon J. E. (1975) Nutritional individuality. In M. Katz, G. T. Keusch and L. J. Mata (eds.) *Interactions of Nutrition and Infection*. Guatemala City, Panama: American Journal of Diseases in Childhood, pp. 422–424.

Gordon J. E., Chitkara I. D. and Wyon J. B. (1963) Weanling diarrhea. *American Journal of the Medical Sciences* 245:345–377.

Gowing L. (1997) Secret births and infanticide in seventeenth-century England. *Past and Present* 156:87–115.

Gowland R. (2001) Playing dead: implications of mortuary evidence for the social construction of childhood in Roman Britain. In G. Davies, A. Gardner and K. Lockyear (eds.) *Proceedings of the 10th Annual Theoretical Roman Archaeology*

Conference, University College London, April 2000. Oxford, UK: Oxbow, pp. 152–168.

Gowland R. (2002) Age as an aspect of social identity in fourth to sixth century A.D. England: the archaeological funerary evidence. Ph.D. thesis, University of Durham, Durham, UK.

Gowland R. and Chamberlain A. (2002) A Bayesian approach to ageing perinatal skeletal material from archaeological sites: implications for the evidence for infanticide in Roman Britain. *Journal of Archaeological Science* 29:677–685.

Graham A. H. and Davies S. M. (1993) *Excavations in Trowbridge, Wiltshire 1977–88: The Prehistoric, Saxon and Saxo-Norman Settlements and the Anarchy Period Castle.* London: Oxford Books.

Grauer A. L. (1991) Life patterns of women from medieval York. In D. Walde and N. D. Willows (eds.) *Proceedings of the 22nd Annual Charcmool Conference,* Charcmool, Canada: Archaeological Association of the University of Calgary, pp. 407–413.

Grauer A. L. (1993) Patterns of anemia and infection from medieval York, England. *American Journal of Physical Anthropology* 91:203–213.

Graunt J. (1662) *Natural and Political Observations.* London: T. Roycroft.

Greene P., Chisick M. and Aaron G. (1994) A comparison of oral health status and need for dental care between abused/neglected children and nonabused/non-neglected children. *Pediatric Dentistry* 16:41–45.

Gresham E. (1975) Birth trauma. *Pediatric Clinics of North America* 22:317–328.

Greulich W. (1976) Some secular changes in the growth of American-born and native Japanese children. *American Journal of Physical Anthropology* 45:553–568.

Greulich W. and Pyle S. (1959) *Radiographic Atlas of Skeletal Development of the Hand and Wrist.* Stanford, CA: Stanford University Press.

Griffith J. P. C. (1919) *The Diseases of Infants and Children.* London: W. B. Saunders.

Grolleau-Raoux J.-L., Crubezy E., Rouge D., Brugne J.-F. and Saunders S. (1997) Harris lines: a study of age-associated bias in counting and interpretation. *American Journal of Physical Anthropology* 103:209–217.

Gruspier K. (1989) A case of rickets in a medieval Jordanian infant. Paper presented at the *16th Annual Meeting of the Paleopathology Association,* San Diego, CA.

Guatelli-Steinberg D. (2003a) Analysis and significance of linear enamel hypoplasia in Plio-Pleistocene hominins. *American Journal of Physical Anthropology* 123:199–215.

Guatelli-Steinberg D. (2003b) Macroscopic and microscopic analyses of linear enamel hypoplasia in Plio-Pleistocene South African hominins with respect to aspects of enamel development and morphology. *American Journal of Physical Anthropology* 120:309–322.

Gunst K., Mesotten K., Carbonez A. and Willems G. (2003) Third molar root development in relation to chronological age: a large sample sized retrospective study. *Forensic Science International* 136:52–57.

Gustafson G. and Koch G. (1974) Age estimation up to 16 years of age based on dental development. *Orthodontic Review* 25:297–306.

Guy H., Masset C. and Baud C.-A. (1997) Infant taphonomy. *International Journal of Osteoarchaeology* 7:221–229.

Haas J. D. (1990) Mortality and morbidity consequences of variation in early child growth. In A. C. Swedlund and G. J. Armelagos (eds.) *Diseases in Populations in Transition: Anthropological and Epidemiological Perspectives.* New York: Bergin & Garvey, pp. 223–247.

Hackett C. J. (1981) Development of caries sicca in a dry calvaria. *Virchow's Archives A (Pathological Anatomy)* **391**:53–79.

Hagen A. (1995) *A Second Handbook of Anglo-Saxon Food and Drink: Production and Distribution.* Chippenham, UK: Anglo-Saxon Books.

Hägg U. and Taranger J. (1982) Maturation indicators and the pubertal growth spurt. *American Journal of Orthodontics* **82**:299–309.

Haglund W. (1997) Scattered skeletal human remains: search strategy considerations for locating missing teeth. In W. Haglund and M. Sorg (eds.) *Forensic Taphonomy: The Postmortem Fate of Human Remains.* New York: CRC Press, pp. 383–394.

Halcomb S. M. C. and Konigsberg L. M. (1995) Statistical study of sexual dimorphism in the human fetal sciatic notch. *American Journal of Physical Anthropology* **97**:113–126.

Hamill P., Drizd T., Johnson C., Reed R. and Roche A. (1977) NCHS growth curves for children birth–18 years. *United States Vital Health Statistics* **165**:1–74.

Hammond G. and Hammond N. (1981) Child's play: a distorting factor in archaeological distribution. *American Antiquity* **46**:634–636.

Hanawalt B. (1986) *The Ties that Bound: Peasant Families in Medieval England.* Oxford, UK: Oxford University Press.

Hanawalt B. (1993) *Growing Up in Medieval London.* Oxford, UK: Oxford University Press.

Hanawalt B. (2002) Medievalists and the study of childhood. *Speculum* **77**:440–460.

Hanihara K. (1961) Criteria for classification of crown characters of the human deciduous dentition. *Journal of the Anthropological Society of Nippon* **69**:27–45.

Hansen L. A. and Winberg J. (1972) Breast milk and defence against infection in the newborn. *Archives of Diseases in Childhood* **47**:845–847.

Hardy A. (1992) Rickets and the rest: child-care, diet and the infectious children's diseases, 1850–1914. *Social History of Medicine* **5**:389–412.

Harila V., Heikkinen T. and Alvesalo L. (2003) Deciduous tooth crown size in prematurely born children. *Early Human Growth and Development* **75**:9–20.

Harila-Kaera V., Heikkinen T., Alvesalo L. and Osborne R. (2001) Permanent tooth crown dimensions in prematurely born children. *Early Human Growth and Development* **62**:131–147.

Harman N. (1917) *Staying the Plague.* London: Methuen.

Harris E., Hicks J. and Barcroft B. (2001) Tissue contributions to sex and race: differences in tooth crown size of deciduous molars. *American Journal of Physical Anthropology* **115**:223–237.

Harris E. and McKee J. (1990) Tooth mineralisation standards for Blacks and Whites from the middle southern United States. *Journal of Forensic Sciences* **35**:859–872.

Harris H. A. (1931) Lines of arrested growth in the long bones in childhood. *British Journal of Radiology* **18**:622–640.

Harris H. A. (1933) *Bone Growth in Health and Disease.* London: Oxford University Press.

Hart N. (1998) Beyond infant mortality: gender and stillbirth in reproductive mortality before the twentieth century. *Population Studies* **52**:215–229.

Harvey R. (2003) Child soldiers. *Childright* **197**:6–8.

Harvey W. (1968) Some dental and social conditions of 1696–1852 connected with St. Bride's church, Fleet Street, London. *Medical History* **12**:62–75.

Hasselwander A. (1902) Untersuchungen über die ossification des menschlichen fussskeletts. *Zeitschrift für Morphologie und Anthropologie* **12**:1–140.

Hauspie R., Chrzastek-Spruch H., Verleyen G., Kozlowska M. and Suzsanne C. (1994) Determinates of growth in body length from birth to 6 years of age: a longitudinal study of Dublin children. *International Journal of Anthropology* **9**:202.

Hayes J. T. (1961) Cystic tuberculosis of the proximal tibial metaphysis with associated involvement of the epiphysis and epiphyseal plate. *Journal of Bone and Joint Surgery* **43**-A:560–567.

Hayward A. R. (1978) Development of immune responsiveness. In F. Falkner and J. M. Tanner (eds.) *Human Growth: A Comprehensive Treatise.* New York: Plenum Press, pp. 593–607.

Hedges R. (2002) Bone diagenesis: an overview of processes. *Archaeometry* **44**:319–328.

Heintz N. (1963) Croissance et puberté feminines au Rwanda. *Mémoires de l'Académie Royale des Sciences Naturelles et Médicales* **12**:1–143.

Helmkamp R. and Falk D. (1990) Age- and sex-associated variation in directional asymmetry of rhesus macaque forelimb bones. *American Journal of Physical Anthropology* **83**:211–218.

Henderson J. (1984) Factors determining the state of preservation of human remains. In A. Boddington, A. N. Garland and R. C. Janaway (eds.) *Death, Decay and Reconstruction: Approaches to Archaeology and Forensic Science.* Manchester, UK: Manchester University Press, pp. 43–54.

Henneberg M. (1977) Proportion of dying children in palaeodemographical studies: estimation by guess or by methodical approach. *Przeglad Antropologiczny* **43**:107–113.

Henneberg M. (1997) Secular trends in body size: indicator of general improvement or specific environmental factors? *American Journal of Physical Anthropology* Suppl. **24**:127.

Henneberg M. and Henneberg R. (1994) Treponematosis in an ancient Greek colony of Metaponto, southern Italy, 580–250 BCE. In O. Dutour, G. Pálfi, J. Berato and J.-P. Brun (eds.) *L'Origine de la syphilis en Europe: Avant ou Après 1493?* Paris: Centre Archéologique du Var, Editions Errance, pp. 92–98.

Henneberg M. and Steyn M. (1994) Preliminary report on the paleodemography of the K2 and Mapungubwe populations (South Africa). *Human Biology* **66**:105–120.

Hensinger R. (1998) Complications of fractures in children. In N. Green and M. Swiontkowski (eds.) *Skeletal Trauma in Children.* Philadelphia, PA: W. B. Saunders, pp. 121–147.

Herring D., Saunders S. and Boyce G. (1994) Bones and the burial registers: infant mortality in a 19th century cemetery from Upper Canada. *Council for Northeast Historical Archaeology Journal* **20**:54–70.

Herring D., Saunders S. and Katzenberg M. (1998) Investigating the weaning process in past populations. *American Journal of Physical Anthropology* **105**:425–439.

Hershkovitz I., Rothschild B. and Latimer B. (1997) Recognition of sickle cell anemia in skeletal remains of children. *American Journal of Physical Anthropology* **104**:213–226.

Hershkovitz I., Greenwald C., Latimer B., *et al.* (2002) Serpens endocrania symmetrica (SES): a new term and a possible clue for identifying intrathoracic diseases in skeletal populations. *American Journal of Physical Anthropology* **118**:201–216.

Hewitt D., Westropp C. K. and Acheson R. M. (1955) Oxford Child Health survey effect of childish ailments on skeletal development. *British Journal of Preventive Social Medicine* **9**:179–186.

Hiernaux J. (1964) Weight/height relationship during growth in Africans and Europeans. *Human Biology* **36**:273–293.

Higgins R. L. and Sirianni J. E. (1995) An assessment of health and mortality of nineteenth century Rochester, New York using historic records and the Highland Park skeletal collections. In A. L. Grauer (ed.) *Bodies of Evidence: Reconstructing History through Skeletal Analysis*. New York: Wiley-Liss, pp. 121–136.

Hildes J. A. and Schaefer O. (1973) Health of Igloolik Eskimos and changes with urbanisation. *Journal of Human Evolution* **2**:241–246.

Hillson S. (1996) *Dental Anthropology*. Cambridge, UK: Cambridge University Press.

Hillson S. and Bond S. (1997) Relationship of enamel hypoplasia to the pattern of tooth crown growth: a discussion. *American Journal of Physical Anthropology* **104**:89–103.

Hillson S., Grigson C. and Bond S. (1998) Dental defects of congenital syphilis. *American Journal of Physical Anthropology* **107**:25–40.

Hillson S. W. (1992) Studies of growth in dental tissues. In J. R. Lukacs (ed.) *Culture, Ecology and Dental Anthropology, Journal of Human Ecology, Special Issue*: 7–23.

Himes J., Yarbrough C. and Martorell R. (1977) Estimation of stature in children from radiographically determined metacarpal length. *Journal of Forensic Sciences* **22**:452–456.

Hodges D. C. and Wilkinson R. G. (1990) Effect of tooth size on the ageing and chronology distribution of enamel hypoplastic defects. *American Journal of Human Biology* **2**:553–560.

Hoey H., Tanner J. and Cox L. (1987) Clinical growth standards for Irish children. *Acta Paediatrica Scandinavica* **338**:1–31.

Hoffman J. M. (1979) Age estimations from diaphyseal lengths: two months to twelve years. *Journal of Forensic Sciences* **24**:461–469.

Holck P. (2001) Specific problems associated with identification from burned and cremated remains. Paper presented at the *First Meeting of the British Association of Human Identification*, University of Glasgow, 2001.

Holcomb S. and Konigsberg L. (1995) Statistical study of sexual dimorphism in the human fetal sciatic notch. *American Journal of Physical Anthropology* **97**:113–125.

Holland T. D. and O'Brien M. J. (1997) Parasites, porotic hyperostosis and the implications of changing perspectives. *American Antiquity* **62**:183–193.

Holman D. J. and Jones R. E. (1998) Longitudinal analysis of deciduous tooth emergence. II. Parametric survival analysis in Bangladeshi, Guatemalan, Japanese, and
Javanese children. *American Journal of Physical Anthropology* **105**:209–230.

Holman D. and Yamaguchi K. (2004) Longitudinal analysis of deciduous tooth emergence. IV. Covariate effects in Japanese children. *American Journal of Physical
Anthropology* **126**:352–358.

Holt J. F. (1950) Vitamin D resistant rickets (refractory rickets). *American Journal of
Roentgenology* **64**:590–602.

Holt L. (1909) *Diseases of Infancy and Childhood*. London: D. Appleton.

Hooper B. (1991) Anatomical considerations. In B. Cunliffe and C. Poole (eds.) *Dane-
bury, An Iron Age Hillfort in Hampshire*, vol. 5, *The Excavations, 1979–88: The
Finds*. London: Council for British Archaeology, pp. 425–431.

Hooper B. (1996) A medieval depiction of infant-feeding in Winchester Cathedral.
Medieval Archaeology **40**:230–233.

Hoppa, R. (1992) Evaluating human skeletal growth: an Anglo-Saxon example. *International Journal of Osteoarchaeology* **2**:275–288.

Hoppa R. and Fitzgerald C. (eds.) (1999) *Human Growth in the Past: Studies from Bones
and Teeth*. Cambridge, UK: Cambridge University Press.

Hoppa R. and Gruspier K. L. (1996) Estimating diaphyseal length from fragmentary
subadult skeletal remains: implications for palaeodemography reconstructions of
a southern Ontario ossuary. *American Journal of Physical Anthropology* **100**:341–
354.

Hoppa R. and Saunders S. (1994) The δl method for examining bone growth in juveniles:
a reply. *International Journal of Osteoarchaeology* **4**:261–263.

Horiuchi H., Kaneko S. and Endo T. (1981) An epidemiological study of the relationship
between air pollution and nasal allergy. *Rhinology* **1**:161–167.

Hoshower L. M. (1994) Brief communication: immunologic aspects of human colostrum
and milk – a misinterpretation. *American Journal of Physical Anthropology* **94**:421–
425.

Hrdy S. B. (1994) Fitness tradeoffs in the history and evolution of delegated mothering
with special reference to wet-nursing, abandonment and infanticide. In S. Parmi-
giani and F. vom Saal (eds.) *Infanticide and Parental Care*. Chur, Switzerland:
Harwood Academic Publishers, pp. 3–41.

Huda T. F. J. and Bowman J. E. (1995) Age determination from dental microstructure
in juveniles. *American Journal of Physical Anthropology* **97**:135–150.

Hughes C., Heylings D. J. A. and Power C. (1996) Transverse (Harris) lines in Irish
archaeological remains. *American Journal of Physical Anthropology* **101**:115–131.

Hühne-Osterloh G. and Grupe G. (1989) Causes of infant mortality in the Middle
Ages revealed by chemical and palaeopathological analyses of skeletal remains.
Zeitschrift für Morphologie und Anthropologie **77**:247–258.

Huizinga J. (1982) A Tellem child: a peculiar case of dwarfism in Africa: In G. Haneveld,
W. Perizonius and P. Janessen (eds.) *Proceedings of the 4th European Meeting of
the Paleopathology Association*, Middelburg-Antwerpen, pp. 63–65

Hummert J. R. (1983) Cortical bone growth and dietary stress among subadults from
Nubia's Batn El Hajar. *American Journal of Physical Anthropology* **62**:167–
176.

Hummert J. R. and Van Gerven D. P. (1983) Skeletal growth in a medieval population from Sudanese Nubia. *American Journal of Physical Anthropology* **62**:471–478.

Humphrey L. (1998) Growth patterns in the modern human skeleton. *American Journal of Physical Anthropology* **105**:57–72.

Humphrey L. (2000) Interpretations of the growth of past populations. In J. S. Derevenski (ed.) *Children and Material Culture*. London: Routledge, pp. 193–205.

Humphrey L., Jeffries T. and Dean M. (2004) Investigation of age at weaning using Sr/Ca ratios in human tooth enamel. *American Journal of Physical Anthropology* Suppl. **38**:117.

Hunt D. (1990) Sex determination in the subadult ilia: an indirect test of Weaver's non-metric sexing method. *Journal of Forensic Sciences* **35**:881–885.

Hunt E. and Gleiser I. (1955) The estimation of age and sex of pre-adolescent children. *American Journal of Physical Anthropology* **13**:79–87.

Hunt E. and Hatch J. W. (1981) The estimation of age at death and ages of formation of transverse lines from measurements of human long bones. *American Journal of Physical Anthropology* **54**:461–469.

Hunter J. (1996) Recovering buried remains. In J. Hunter, C. Roberts and A. Martin (eds.) *Studies in Crime: An Introduction to Forensic Archaeology*. London: Batsford, pp. 40–57.

Huss-Ashmore R., Goodman A. and Armelagos G. (1982) Nutritional inference from paleopathology. In B. Schiffer (ed.) *Advances in Archaeological Method and Theory*. New York: Academic Press, pp. 395–474.

Hutchins L. (1998) Standards of infant long bone diaphyseal growth from a late nineteenth century and early twentieth century almshouse cemetery. M.Sc. dissertation, University of Wisconsin–Madison, WI.

Hutchinson J. (1857) On the influence of hereditary syphilis on the teeth. *Transactions of the Odontological Society of Great Britain* **2**:95–106.

Hutchinson J. (1858) Report on the effects of infantile syphilis in marring the development of teeth. *Transactions of the Pathological Society London* **9**:449–456.

Huxley A. (1998) Analysis of shrinkage in human fetal diaphyseal lengths from fresh to dry bone using Petersohn and Köhler's data. *Journal of Forensic Sciences* **43**:423–426.

Huxley A. and Angevine J. (1998) Determination of gestational age from lunar age assessments in human fetal remains. *Journal of Forensic Sciences* **43**:1254–1256.

Huxley A. and Jimenez S. (1996) Technical note: error in Olivier and Pineau's regression formulae for calculation of stature and lunar age from radial diaphyseal length in forensic fetal remains. *American Journal of Physical Anthropology* **100**:435–437.

Huxley A. and Kósa F. (1999) Calculation of percentage shrinkage in human fetal diaphyseal lengths from fresh bone to carbonised and calcined bone using Petersohn and Köhler's data. *Journal of Forensic Sciences* **44**:577–583.

Imrie J. and Wyburn G. (1958) Assessment of age, sex and height from immature human bones. *British Medical Journal* **1**:128–131.

Ingalls N. (1927) Studies on femur. III. Effects of maceration and drying in the White and the Negro. *American Journal of Physical Anthropology* **10**:297–321.

Ingvarsson-Sundström A. (2003) Children lost and found: a bioarchaeological study of Middle Helladic children in Asine with a comparison to Lerna. Ph.D. thesis, Uppsala University, Uppsala, Sweden.

Intelligence and Security Council (2000) *Criminal Statistics England and Wales 1999: Statistics Relating to Crime and Criminal Proceedings for the Year 1999.* London: The Stationery Office.

Iregren E. and Boldsen J. (1994) Patterns of long bone growth in the Medieval Swedish community Westerhaus. *International Journal of Anthropology* **9**:205.

Jackes M. K. (1983) Osteological evidence for smallpox: a possible case from seventeenth century Ontario. *American Journal of Physical Anthropology* **60**:75–81.

Jacobi K. and Danforth M. (2002) Analysis of interobserver error scoring patterns in porotic hyperostosis and cribra orbitalia. *International Journal of Osteoarchaeology* **12**:248–258.

Jacobi K., Cook D., Corruccini R. and Handler J. (1992) Congenital syphilis in the past: slaves at Newton Plantation, Barbados, West Indies. *American Journal of Physical Anthropology* **89**:145–158.

Jakob B. (2003) Prevalence and patterns of disease in early medieval populations: a comparison of skeletal samples from fifth to eighth century AD Britain and southwestern Germany. Ph.D. thesis, University of Durham, Durham, UK.

Jankauskas R. (1999) Tuberculosis in Lithuania: paleopathological and historical correlations. In G. Pálfi, O. Dutour, J. Deák and I. Hutás (eds.) *Tuberculosis: Past and Present.* Szeged, Hungary: Golden Book Publisher Ltd and Tuberculosis Foundation, pp. 551–558.

Jankauskas R. and Schultz M. (1995) Meningeal reactions in a late medieval–early modern child population from Alytus, Lithuania. *Journal of Paleopathology* **7**: 106.

Jantz R. L. and Owsley D. W. (1984) Long bone growth variation among Arikara skeletal populations. *American Journal of Physical Anthropology* **63**:13–20.

Jasuja O., Harbhajan S. and Anupama K. (1997) Estimation of stature from stride length while walking fast. *Forensic Science International* **86**:181–186.

Jelliffe D. and Blackman V. (1962) Bahima disease: possible 'milk anemia' in late childhood. *Tropical Pediatrics* **61**:774–779.

Jelliffe D. and Jelliffe E. (1978) *Human Milk in the Modern World: Psychosocial, Nutritional, and Economic Significance.* Oxford, UK: Oxford University Press.

Johansson S. and Owsley D. (2002) Welfare history on the Great Plains: mortality and skeletal health 1650–1900. In R. H. Steckel and J. C. Rose (eds.) *The Backbone of History: Health and Nutrition in the Western Hemisphere.* Cambridge, UK: Cambridge University Press, pp. 524–562.

Johnston F. E. (1962) Growth of the long bones of infants and young children at Indian Knoll. *American Journal of Physical Anthropology* **20**:249–254.

Johnston F. E. (1968) Growth of the skeleton in earlier peoples. In D. R. Brothwell (ed.) *The Skeletal Biology of Past Human Populations.* London: Pergamon Press, pp. 57–66.

Johnston F. E. (1969) Approaches to the study of developmental variability in human skeletal populations. *American Journal of Physical Anthropology* **31**:335–342.

Johnston F. E. and Snow C. (1961) The reassessment of the age and sex of the Indian Knoll skeletal population: demographic and methodological aspects. *American Journal of Physical Anthropology* **19**:237–244.

Johnston F. E and Zimmer L. O. (1989) Assessment of growth and age in the immature skeleton. In M. Y. Iscan and K. A. R. Kennedy (eds.) *Reconstruction of Life from the Skeleton*. New York: Alan R. Liss, pp. 11–21.

Johnston F. E., Borden M. and MacVean R. B. (1975) The effects of genetic and environmental factors upon the growth of children in Guatemala City. In E. S. Watts, F. E. Johnston and G. W. Lasker (eds.) *Biosocial Interrelations in Population Adaptation*. Paris: Mouton, pp. 377–388.

Jones E. and Ubelaker D. (2001) Demographic analysis of the Voegtly cemetery sample, Pittsburgh, Pennsylvania. *American Journal of Physical Anthropology* Suppl. **32**:86.

Jones M., James D., Cory C., Leadbeatter S. and Nokes L. (2003) Subdural haemorrhage sustained in a baby-rocker? A biomechanical approach to causation. *Forensic Science International* **131**:14–21.

Kahana T., Birkby W., Goldin L. and Hiss J. (2003) Estimation of age in adolescents: the basilaris synchondrosis. *Journal of Forensic Sciences* **48**:1–5.

Kamp K. (2001) Where have all the children gone? The archaeology of childhood. *Journal of Archaeological Method and Theory* **8**:1–34.

Karlberg J. (1998) The human growth curve. In S. J. Ulijaszek, F. C. Johnston and M. A. Preece (eds.) *The Cambridge Encyclopedia of Human Growth and Development*. Cambridge, UK: Cambridge University Press, pp. 108–115.

Karlberg P., Taranger J., Engstrom I., Lichtenstein H. and Svennberg-Redegren I. (1976) *The Somatic Development of Children in a Swedish Urban Community*. Göteborg, Sweden: University of Göteborg.

Katzenberg A. (2000) Stable isotope analysis: a tool for studying past diet, demography, and life history. In A. Katzenberg and S. R. Saunders (eds.) *Biological Anthropology of the Human Skeleton*. New York: Wiley-Liss, pp. 305–327.

Katzenberg M. A. and Lovell N. C. (1999) Stable isotope variation in pathological bone. *International Journal of Osteoarchaeology* **9**:316–324.

Katzenberg M. A. and Pfeiffer S. (1995) Nitrogen isotope evidence for weaning age in a nineteenth century Canadian skeletal sample. In A. L. Grauer (ed.) *Bodies of Evidence: Reconstructing History through Skeletal Analysis*. New York: Wiley-Liss, pp. 221–235.

Katzenberg M. A., Saunders S. R. and Fitzgerald W. R. (1993) Age differences in stable carbon and nitrogen isotope ratios in a population of prehistoric maize horticulturists. *American Journal of Physical Anthropology* **90**:267–281.

Katzenberg A., Herring A. and Saunders S. (1996) Weaning and infant mortality: evaluating the skeletal evidence. *Yearbook of Physical Anthropology* **39**:177–199.

Katzenberg A., Oetelaar G., Oetelaar J., *et al.* (2005) Identification of historical human skeletal remains: a case study using skeletal and dental age, history and DNA. *International Journal of Osteoarchaeology* **15**:61–72.

Keene H. (1998) Are we underestimating canine sexual dimorphism in humans? *American Journal of Physical Anthropology* Suppl. **26**:197.

Keipert J. A. and Campbell P. E. (1970) Recurrent hyperostosis of the clavicles: an undiagnosed syndrome. *Australian Paediatric Journal* **6**:97–104.

Kelley M. and Micozzi M. (1984) Rib lesions in chronic pulmonary tuberculosis. *American Journal of Physical Anthropology* **65**:381–387.

Kelly A., Shaw N., Thomas A., Pynsent P. and Baker D. (1997) Growth of Pakistani children in relation to the 1990 growth standards. *Archives of Diseases in Childhood* **77**:401–405.

Kempe C., Silverman F., Steele B., Droegemueller W. and Silver H. (1962) The battered-child syndrome. *Journal of the American Medical Association* **181**:105–112.

Kennedy G. (2005) From the ape's to the weanling's dilemma: early weaning and its evolutionary context. *Journal of Human Evolution* **48**:123–145.

Kent S. (1986) The influence of sedentism and aggregation on porotic hyperostosis and anemia: a case study. *Man* **21**:605–636.

Kerley E. (1976) Forensic anthropology and crimes involving children. *Journal of Forensic Sciences* **21**:333–339.

Kerley E. (1978) The identification of battered-infant skeletons. *Journal of Forensic Sciences* **23**:163–168.

Key C. (2000) The evolution of human life history. *World Archaeology* **31**:329–350.

Kilgore P. E., Holman R. C., Clarke M. J. and Glass R. I. (1995) Trends of diarrheal disease-associated mortality in US children, 1968 through 1991. *Journal of the American Medical Association* **274**:1143–1148.

King J. (2001) Effect of reproduction on the bioavailability of calcium, zinc and selenium. *Journal of Nutrition* **131**:1355–1358.

King S. (1997) Dying with style: infant death and its context in a rural industrial township, 1650–1830. *Social History of Medicine* **10**:3–24.

King S. E. and Ulijaszek S. J. (2000) Invisible insults during growth and development: contemporary theories and past populations. In R. D. Hoppa and C. M. Fitzgerald (eds.) *Human Growth in the Past: Studies from Bones and Teeth*. Cambridge, UK: Cambridge University Press, pp. 161–182.

King T., Humphrey L. and Hillson S. (2002) Developmental stress in a post-medieval population of Londoners. *American Journal of Physical Anthropology* Suppl. **34**:95.

Kipel K. (ed.) (2003) *The Cambridge Historical Dictionary of Disease*. Cambridge, UK: Cambridge University Press.

Kleinman P. (1987) Skeletal trauma: general considerations. In P. Kleinman (ed.) *Diagnostic Imaging of Child Abuse*. Baltimore, MD: Williams & Wilkins, pp. 5–27.

Kleinman P., Marks S., Richmond J. and Blackbourne B. (1995) Inflicted skeletal injury: a postmortem radiologic–histopathologic study in 31 infants. *American Journal of Radiology* **165**:647–650.

Kleinman P. and Schlesinger A. (1997) Mechanical factors associated with posterior rib fractures: laboratory and case studies. *Pediatric Radiology* **27**:87–91.

Knick S. G. (1982) Linear enamel hypoplasia and tuberculosis in pre-Columbian North America. *Ossa* **8**:131–138.

Knight B. (1986) The history of child abuse. *Forensic Science International* **30**:135–141.

Knight B. (1996) *Forensic Pathology*, 2nd edn. London: Arnold.

Knight B. and Whittaker D. (1997) Medical and dental investigations in the Rosemary West case. *Medico-Legal Journal* **65**:107–121.

Knodel J. and Kintner H. (1977) The impact of breast feeding patterns on the biometric analysis of infant mortality. *Demography* **14**:391–409.

Knott N. (1967) Identification by the teeth of casualties in the Aberfan disaster. *British Dental Journal* **122**:144–145.

Koganei H. (1912) Cribra cranii und cribra orbitalia. *Mitt MedFak Toyko* **10**:113–154.

Kohlberg L. (1966) A cognitive-developmental analysis of children's sex-role concepts and attitudes. In C. Maccoby (ed.) *The Development of Sex Differences*. Stanford, CA: Stanford University Press, pp. 80–173.

Konigsberg L., Frankenberg S. and Walker R. (1997) Regress what on what? Palaeodemographic age estimation as a calibration problem. In R. R. Paine (ed.) *Integrating Archaeological Demography: Multidisciplinary Approaches to Prehistoric Populations*. Carbondale, IL: Center for Archaeological Investigations, Southern Illinois University, pp. 64–88.

Konigsberg L. and Frankenberg S. (1992) Estimation of age structure in anthropological demography. *American Journal of Physical Anthropology* **89**:235–256.

Koren G. (1995) Measurement of drugs in neonatal hair: a window to fetal exposure. *Forensic Science International* **70**:77–82.

Kósa F. (1989) Age estimation from the fetal skeleton. In M. Y. Iscan (ed.) *Age Markers in the Human Skeleton*. Springfield, IL: Charles C. Thomas, pp. 21–54.

Kósa F. and Castellana C. (2005) New forensic anthropological approachment for the age determination of human fetal skeletons on the base of the morphometry of vertebral column. *Forensic Science International* **147**:69–74.

Kovacs C. (2003) Fetal mineral homeostasis. In F. Glorieux, J. Pettifor and M. Jüppner (eds.) *Pediatric Bone: Biology and Diseases*. New York: Academic Press, pp. 271–302.

Koziel S. and Ulijaszek S. (2001) Waiting for Travers and Willard: do the rich really favor sons? *American Journal of Physical Anthropology* **115**:71–79.

Kraus B. S. and Jordan R. E. (1965) *The Human Dentition before Birth*. London: Henry Kimpton.

Krause J. T. (1969) English population movements between 1700–1850. In M. Drake (ed.) *Population in Industrialization*. London: Methuen, pp. 118–127.

Kreshover S. J. (1960) Metabolic disturbances in tooth formation. *Annals of the New York Academy of Sciences* **85**:161–167.

Kreutz K., Teichmann G. and Schultz M. (1995) Palaeoepidemiology of inflammatory processes of the skull: a comparative study of two early medieval infant populations. *Journal of Paleopathology* **7**:108.

Kronfield R. and Schour I. (1939) Neonatal dental hypoplasia. *Journal of the American Dental Association* **26**:18–32.

Kuefler M. S. (1991) A wryed existence: attitudes towards children in Anglo-Saxon England. *Journal of Social History* **23**:823–834.

Kullman L. (1995) Accuracy of two dental and one skeletal age estimation method in Swedish adolescents. *Forensic Science International* **75**:225–236.

Kurniewicz-Witczakowa R., Miesowicz I., Niedzwiecka Z. and Pietrzak M. (1983) *Rozwoj Fizycyny Dzieci i Mlodzieizy Warszawskiej*. Warsaw: Institute of Mother and Child.

Kuzawa C. (1998) Adipose tissue in human infancy and childhood: an evolutionary perspective. *American Journal of Physical Anthropology* **41**:177–209.

Lagier R., Kramar C. and Baud C. A. (1987) Femoral unicameral bone cyst in a medieval child: radiological and pathological study. *Pediatric Radiology* **17**:498–500.

Lallo J. W. (1973) The skeletal biology of three prehistoric American Indian societies from Dickson Mounds. Ph.D. thesis, University of Massachusetts, Amherst, MA.

Lallo J. W., Armelagos J. G. and Mensforth R. P. (1977) The role of diet, disease and physiology in the origin of porotic hyperostosis. *Human Biology* **49**:471–483.

Lambert P. (2002) Rib lesions in a prehistoric Puebloan sample from southwestern Colorado. *American Journal of Physical Anthropology* **117**:281–292.

Lampl M. (2005) Grandma's right: a sleeping baby may be a growing baby. *American Journal of Physical Anthropology* Suppl. **40**:134.

Lampl M. and Jeanty P. (2003) Timing is everything: a reconsideration of fetal growth velocity patterns identifies the importance of individual and sex differences. *American Journal of Human Biology* **15**:667–680.

Lampl M. and Johnston F. E. (1996) Problems in the aging of skeletal juveniles: perspectives from maturation assessments of living children. *American Journal of Physical Anthropology* **101**:345–355.

Lampl M., Veldhuis J. D. and Johnson M. L. (1992) Salutation and stasis: a model of human growth. *Science* **258**:801–803.

Lampl M., Kuzawa C. and Jeanty P. (2003) Prenatal smoke exposure alters growth in limb proportions and head shape in the midgestation human fetus. *American Journal of Human Biology* **15**:533–546.

Lane J. (2001) *Midwifery and Nursing*. London: Routledge.

Langer W. L. (1974) Infanticide: a historical survey. *History of Childhood Quarterly* **1**:129–134.

Lanphear K. M. (1990) Frequency and distribution of enamel hypoplasias in a historic skeletal sample. *American Journal of Physical Anthropology* **81**:35–43.

Lanzkowsky P. (1968) Radiological features of iron-deficiency anemia. *American Journal of Diseases in Children* **116**:16–29.

Lanzkowsky P. (1977) Osseous changes in iron deficiency anemia: implications for paleopathology. In E. Cockburn (ed.) *Porotic Hyperostosis: An Enquiry*. Detroit, MI: Paleopathology Association, pp. 23–34.

Larsen C. S. (1997) *Bioarcheology: Interpreting Behavior from the Human Skeleton*. Cambridge, UK: Cambridge University Press.

Lease L. and Scuilli P. (2005) Brief communication: discrimination between European-American and African-American children based on deciduous dental metrics and morphology. *American Journal of Physical Anthropology* **126**:56–60.

Lee K. (1994) Attitudes and prejudices towards infanticide: Carthage, Rome and today. *Archaeological Review from Cambridge* **13**:65–79.

Leidy Sievert L. (2003) Monitored growth: anthropometrics and health history records at a private New England middle school, 1935–1960. In H. Herring and A. Swedlund (eds.) *Human Biologists in the Archives*. Cambridge, UK: Cambridge University Press, pp. 130–158.

Levine R. and Keen J. (1974) Neonatal enamel hypoplasia in association with symptomatic neonatal hypocalcaemia. *British Dental Journal* **137**:429–433.

Lewis B. (1998) Prehistoric juvenile rheumatoid arthritis in a Precontact Louisiana Native population reconsidered. *American Journal of Physical Anthropology* **106**:229–248.

Lewis M. E. (1999) The impact of urbanisation and industrialisation in medieval and post-medieval Britain: an assessment of the morbidity and mortality of non-adult skeletons from the cemeteries of two urban and two rural sites in England (AD 850–1859). Ph.D. thesis, University of Bradford, Bradford, UK.

Lewis M. E. (2000) Non-adult palaeopathology: current status and future potential. In M. Cox and S. Mays (eds.) *Human Osteology in Archaeology and Forensic Science*. London: Greenwich Medical Media Ltd, pp. 39–57.

Lewis M. E. (2002a) The impact of industrialisation: comparative study of child health in four sites from medieval and post-medieval England (AD 850–1859). *American Journal of Physical Anthropology* **119**:211–223.

Lewis M. E. (2002b) Infant and childhood leprosy: clinical and palaeopathological implications. In C. Roberts, M. Lewis and K. Manchester (eds.) *The Past and Present of Leprosy*, BAR (International Series) no. S1054. Oxford, UK: Archaeopress, pp. 163–170.

Lewis M. E. (2002c) *Urbanisation and Child Health in Medieval and Post-Medieval England*, BAR (British Series) no. 339. Oxford, UK: Archaeopress.

Lewis M. E. (2004) Endocranial lesions in non-adult skeletons: understanding their aetiology. *International Journal of Osteoarchaeology* **14**:82–97.

Lewis M. E. and Gowland R. (2005) Infantile cortical hyperostosis: cases, causes, constraints. Paper presented at the *32nd Annual Meeting of the Paleopathology Association*, 5–6 April 2005, Milwaukee, WI, USA.

Lewis M. E. and Roberts C. A. (1997) Growing pains: the interpretation of stress indicators. *International Journal of Osteoarchaeology* **7**:581–586.

Lewis M. E. and Rutty G. (2003) The endangered child: the personal identification of children in forensic anthropology. *Science and Justice* **43**:201–209.

Lewis M. E., Roberts C. A. and Manchester K. (1995) Comparative study of the prevalence of maxillary sinusitis in later medieval urban and rural populations in northern England. *American Journal of Physical Anthropology* **98**:497–506.

Lillehammer G. (1989) A child is born: the child's world in an archaeological perspective. *Norwegian Archaeological Review* **22**:89–105.

Lilley S. M., Stroud G., Brothwell D. R. and Williamson M. H. (eds.) (1994) *The Jewish Burial Ground at Jewbury*. York, UK: Council for British Archaeology.

Lincoln E. M. and Sewell E. M. (1963) *Tuberculosis in Children*. New York: McGraw-Hill.

Liston M. and Papadopoulos J. (2004) The "Rich Athenian Lady" was pregnant: the anthropology of a geometric tomb reconsidered. *Hesperia* **73**:7–38.

Liversidge H. M. and Molleson T. I. (1995) Spitalfields children: the influence of rickets on dental formation in early childhood. *Bone* **16**:693–694.

Liversidge H. M. and Molleson T. (1999) Developing permanent tooth length as an estimate of age. *Journal of Forensic Sciences* **44**:917–920.

Liversidge H., Lyons F. and Hector M. (2003) The accuracy of three methods of age estimation using radiographic measurements of developing teeth. *Forensic Science International* **131**:22–29.

Lleonart R., Riego E., Suárez R., Ruiz R. and de la Fuente J. (1999) Analyses of DNA from ancient bones of a pre-Columbian Cuban woman and a child. *Genetics and Molecular Biology* 22:285–289.

Logan W. H. G. and Kronfield R. (1933) Development of the human jaws and surrounding structures from birth to the age of fifteen years. *Journal of the American Dental Association* 20:379–427.

Lomax E. (1979) Infantile syphilis as an example of nineteenth century belief in the inheritance of acquired characteristics. *Journal of the History of Medicine* 34:23–39.

Lomax E. (1986) Difficulties in diagnosing infantile scurvy before 1878. *Medical History* 30:70–80.

Lorber J. (1958) Intracranial calcifications following tuberculosis meningitis in children. *Acta Radiology* 50:204–210.

Loth S. (1996) *Sexual Dimorphism in the Human Mandible: An Evolutionary and Developmental Perspective*. Johannesburg, South Africa: University of Witwatersrand.

Loth S. and Henneberg M. (1996) Mandibular ramus flexure: a new morphologic indicator of sexual dimorphism in the human skeleton. *American Journal of Physical Anthropology* 99:473–485.

Loth S. and Henneberg M. (2001) Sexually dimorphic mandibular morphology in the first few years of life. *American Journal of Physical Anthropology* 115:179–186.

Loudon I. (1992) *Death in Childbirth: An International Study of Maternal Care and Maternal Mortality 1800–1950*. Oxford, UK: Clarendon Press.

Lovejoy C. O., Russell K. F. and Harrison M. L. (1990) Long bone growth velocity in the Libben population. *American Journal of Human Biology* 2:533–541.

Lovell N. and Whyte I. (1999) Patterns of dental enamel defects at Ancient Mendes, Egypt. *American Journal of Physical Anthropology* 110:69–80.

Lovey H. (1983) Maturation of permanent teeth in Black and Latino children. *Acta de Odontologia Pediatrica* 4:59–62.

Lucy S. (1994) Children in early medieval cemeteries. *Archaeological Review from Cambridge* 13:21–34.

Lucy D., Aykroyd R., Pollard A. and Solheim T. (1996) A Bayesian approach to adult human age estimation from dental observations by Johanson's age changes. *Journal of Forensic Sciences* 41:189–194.

Ludloff K. (1903) Über Wachstum und Architektur der unteren Femurepiphyse und oberen Tibiaepiphyse: Ein Beitrag zur Röntgendiagnostik. *Bruns Beitrag Klinische Chirurgie* 38:64–75.

Lukacs J. and Walimbe S. (1998) Physiological stress in prehistoric India: new data on localised hypoplasia on primary canines linked to climate and subsistence change. *Journal of Archaeological Science* 25:571–585.

Lukacs J., Nelson G. and Walimbe S. (2001) Enamel hypoplasia and childhood stress in prehistory: new data from India and Southwest Asia. *Journal of Archaeological Science* 28:1159–1169.

Lunt D. A. (1972) The dentition in a group of mediaeval Scottish children. *British Dental Journal* 132:443–446.

Lunt R. C. and Law D. B. (1974) A review of the chronology of calcification of deciduous teeth. *Journal of the American Dental Association* 89:599–606.

Lynch M. (1985) Child abuse before Kempe: an historical literature review. *Child Abuse and Neglect* **9**:7–15.

Lythgoe A. (1965) *The Predynastic Cemetery N7000, Naga-ed-Der*. Berkeley, CA: University of California Press.

Maat G. (1982) Scurvy in Dutch whalers buried in Spitsbergen. In G. Haneveld, W. Perizonius and P. Janessen (eds.) *Proceedings of the 4th European Meeting of the Paleopathology Association*, Middelburg-Antwerpen, pp. 82–93.

Maat G. (1984) Dating and rating of Harris's lines. *American Journal of Physical Anthropology* **63**:291–299.

Maat G. (2004) Scurvy in adults and youngsters: the Dutch experience. – a review of the history and pathology of a disregarded disease. *International Journal of Osteoarchaeology* **14**:77–81.

Maat G. and Uytterschaut H. (1984) Microscopic observations on scurvy in Dutch whalers buried at Spitsbergen. In V. Capecchi and M. Rabino (eds.) *Proceedings of the 6th European Meeting of the Paleopathology Association*. Siena, Italy: Siena University Press, pp. 211–218.

Macchiarelli R., Bondioli L., Censi L., *et al.* (1994) Intra- and interobserver concordance in scoring Harris lines: a test on bone sections and radiographs. *American Journal of Physical Anthropology* **95**:77–83.

MacCurdy G. (1923) Human skeletal remains from the highlands of Peru. *American Journal of Physical Anthropology* **6**:217–352.

Macko S., Engel M., Andrusevich V., *et al.* (1999) Documenting the diet in ancient human populations through stable isotope analysis of hair. *Philosophical Transactions of the Royal Society of London* **354**:65–76.

Macpherson P. and Chenery C. (2004) Weaning in early medieval England. *American Journal of Physical Anthropology* Suppl. **38**:140.

Magennis A. L. (1990) Growth velocity as a factor influencing the formation of transverse lines. *American Journal of Physical Anthropology* Suppl. **81**:262.

Mahler P. (1968) Growth of long bones in a prehistoric population from Sudanese Nubia. M.A. dissertation, University of Utah, Salt Lake City, UT.

Malgosa A., Alesan A., Safont S., Ballbé M. and Ayala M. (2004) A dystotic childbirth in the Spanish Bronze Age. *International Journal of Osteoarchaeology* **14**:98–103.

Mankin H. J. (1974) Rickets, osteomalacia and renal osteodystrophy. *Journal of Bone and Joint Surgery* **56**-A:101–128 and 352–386.

Mansilla J. and Pijoan C. (1995) Brief communication: a case of congenital syphilis during the Colonial Period in Mexico City. *American Journal of Physical Anthropology* **97**:187–195.

Mansilla J., Solis C., Chávez-Lomeli M. and Garna J. (2003) Analysis of colored teeth from Pre-Columbian Tlatelolco: postmortem transformations or intravitam processes. *American Journal of Physical Anthropology* **120**:73–82.

Manwaring M. (2002) Infant burial in early medieval Wales. M.Sc. dissertation, University of Wales Cardiff, Cardiff, UK.

Maresh M. M. (1955) Linear growth of long bones of extremities from infancy through adolescence. *American Journal of Diseases in Children* **89**:725–742.

Maresh M. M. and Deming J. (1939) The growth of the long bones in 80 infants' roentgenograms versus anthropometry. *Child Development* **10**:91–106.

Markland A. (2003) Analysis of strontium/calcium ratios and zinc concentrations as indicators of pregnancy and lactation in archaeological human and pig remains. M.Sc. dissertation, Bournemouth University, Bournemouth, UK.

Marks M., Bennett J. and Wilson O. (1997) Digital video image capture in establishing positive identification. *Journal of Forensic Sciences* **42**:120–124.

Marshall W. A. (1968) Problems in relating the presence of transverse lines in the radius to the occurrence of disease. In D. R. Brothwell (ed.) *The Skeletal Biology of Past Human Populations*. London: Pergamon Press, pp. 245–261.

Martorell R., Khan K. L. and Schroeder D. G. (1994) Reversibility of stunting: epidemiological findings in children from developing countries. *European Journal of Clinical Nutrition* **48**:S45–S57.

Masali M. and Chiarelli B. (1969) Demographic data on the remains of ancient Egyptians. *Journal of Human Evolution* **1**:161–169.

Massler M. and Schour I. (1946) The appositional life span of the enamel and dentine forming cells. I. Human deciduous teeth and first permanent molars. *Journal of Dental Research* **25**:145–150.

Mathai M., Scramm M., Baravilara W., *et al.* (2004) Ethnicity and fetal growth in Fiji, Australian and New Zealand. *Journal of Obstetrics and Gynaecology* **44**:318–321.

Mays S. (1985) The relationship between Harris line formation and bone growth and development. *Journal of Archaeological Science* **12**:207–220.

Mays S. (1993) Infanticide in Roman Britain. *Antiquity* **67**:883–888.

Mays S. (1995) The relationship between Harris lines and other aspects of skeletal development in adults and juveniles. *Journal of Archaeological Science* **22**:511–520.

Mays S. (2000a) The archaeology and history of infanticide, and its occurrence in earlier populations. In J. Sofaer Derevenski (ed.) *Children and Material Culture*. London: Routledge, pp. 180–190.

Mays S. (2000b) Linear and appositional long bone growth in earlier human populations: a case study from medieval England. In R. D. Hoppa and C. M. Fitzgerald (eds.) *Human Growth in the Past: Studies from Bones and Teeth*. Cambridge, UK: Cambridge University Press, pp. 290–312.

Mays S. (2003) Comment on 'A Bayesian approach to aging perinatal skeletal material from archaeological sites: implications for the evidence for infanticide in Roman Britain' by R. L. Gowland and A. T. Chamberlain. *Journal of Archaeological Science* **30**:1695–1700.

Mays S. and Cox M. (2000) Sex determination in skeletal remains. In M. Cox and S. Mays (eds.) *Human Osteology in Archaeology and Forensic Science*. London: Greenwich Medical Media Ltd, pp. 117–130.

Mays S. and Faerman M. (2001) Sex identification in some putative infanticide victims from Roman Britain using ancient DNA. *Journal of Archaeological Science* **28**:555–559.

McCollum E., Simmonds N., Becker J. and Shipley P. (1922) An experimental demonstration of the existence of a vitamin which promotes calcium deposition. *Journal of Biological Chemistry* **53**:293–298.

McHenry H. (1968) Transverse lines in long bones of prehistoric California Indians. *American Journal of Physical Anthropology* **29**:1–18.

McHenry H. M. and Schulz P. D. (1976) The association between Harris lines and enamel hypoplasia in prehistoric California Indians. *American Journal of Physical Anthropology* **44**:507–512.

McNeill W. H. (1979) Historical patterns of migration. *Current Anthropology* **20**:95–102.

Melikian M. and Waldron T. (2003) An examination of skulls from two British sites for possible evidence of scurvy. *International Journal of Osteoarchaeology* **13**:207–212.

Mellanby E. (1919) An experimental investigation on rickets. *Lancet* **1**:407–412.

Mellits E. D., Dorst J. P. and Cheek D. B. (1971) Bone age: its contribution to the prediction of maturational or biological age. *American Journal of Physical Anthropology* **35**:381–384.

Mendlewicz M., Jean-Louise G., Gekker M. and Rapaport M. (1999) Neonaticide in the city of Rio de Janerio: forensic and psycholegal perspectives. *Journal of Forensic Sciences* **44**:741–745.

Mensforth R. P. (1985) Relative tibia long bone growth in the Libben and Bt-5 prehistoric skeletal populations. *American Journal of Physical Anthropology* **68**:247–262.

Mensforth R. P., Lovejoy O. C., Lallo J. W. and Armelagos G. J. (1978) The role of constitutional factors, diet and infectious disease in the etiology of porotic hyperostosis and periosteal reactions in prehistoric infants and children. *Medical Anthropology* **2**:1–59.

Merbs C. (1997) Eskimo skeleton taphonomy with identification of possible polar bear victims. In W. Haglund and M. Sorg (eds.) *Forensic Taphonomy: The Postmortem Fate of Human Remains*. New York: CRC Press, pp. 249–262.

Merchant V. L. and Ubelaker D. H. (1977) Skeletal growth of the protohistoric Arikara. *American Journal of Physical Anthropology* **46**:61–72.

Meredith H. V. (1982) Research between 1950 and 1980 on urban–rural differences in body size and growth rate of children and youths. *Advances in Child Development and Behavior* **17**:83–138.

Meskell L. (1994) Dying young: the experience of death at Deir el Medina. *Archaeological Review from Cambridge* **13**:35–45.

Meskell L. (2000) Cycles of life and death: narrative homology and archaeological realities. *World Archaeology* **31**:423–441.

Metcoff J. (1978) Association of fetal growth with maternal nutrition. In F. Falkner and J. M. Tanner (eds.) *Human Growth: A Comprehensive Treatise*. New York: Plenum Press, pp. 415–460.

Meyer C., Jung C., Kohl T., *et al.* (2002) Syphilis 2001: a palaeopathological reappraisal. *Homo* **53**:39–58.

Michael A. and Brauner P. (2004) Erroneous gender identification by the amelogenin sex test. *Journal of Forensic Sciences* **49**:1–2.

Miles A. E. W. (1963) Dentition in the estimation of age. *Journal of Dental Research* **42**:255–263.

Miles A. E. W. (1989) *An Early Christian Chapel and Burial Ground on the Isle of Ensay, Outer Hebrides, Scotland with a Study of the Skeletal Remains*, BAR (British Series) no. 212. Oxford: Archaeopress.

Miles A. E. W. and Bulman J. S. (1994) Growth curves of immature bones from a Scottish island population of sixteenth to mid-nineteenth century: limb-bone diaphysis and some bones of the hand and foot. *International Journal of Osteoarchaeology* 4:121–136.

Milner G. (1982) Measuring prehistoric levels of health: a study of Mississippian period skeletal remains from the American Bottom, Illinois. Ph.D. thesis, Northwestern University, Evanston, IL.

Milner G., Wood J. and Boldsen J. (2000) Paleodemography. In A. Katzenberg and S. R. Saunders (eds.) *Biological Anthropology of the Human Skeleton*. New York: Wiley-Liss, pp. 467–497.

Mittler D. and Sheridan S. (1992) Sex determination in subadults using auricular surface morphology: a forensic science perspective. *Journal of Forensic Sciences* 37:1068–1075.

Mizoguichi Y. (1985) *Shoveling: A Statistical Analysis of its Morphology*. Tokyo: University of Tokyo Press.

Moggi-Cecchi J., Pacciani E. and Pinto-Cisternas J. (1994) Enamel hypoplasia and age at weaning in 19th century Florence, Italy. *American Journal of Physical Anthropology* 93:299–306.

Mogle P. and Zias J. (1995) Trephination as a possible treatment for scurvy in a middle Bronze Age (ca. 2200 BC) skeleton. *International Journal of Osteoarchaeology* 5:77–81.

Möller V. (1862) Zwei von acuter Rachitis. *Königsberger Medizinische Jahrbücher* 3:136–149.

Møller-Christensen V. (1982) *Æbelholt Kloster*. Copenhagen: Nationalmuseet.

Møller-Christensen V. and Sandison A. T. (1963) Usura orbitae (cribra orbitalia) in the collection of crania in the Anatomy Department of the University of Glasgow. *Pathological Microbiology* 26:175–183.

Molleson T. (1987) Anne Mowbray and the Princes in the Tower: a study in identity. *London Archaeologist* 5:258–262.

Molleson T. (1992) Retardation of growth and early weaning of children in prehistoric populations. *Acta Musei Nationalis Pragae* 46:182–188.

Molleson T. and Cox M. (1988) A neonate with cut bones from Poundbury Camp, 4th century AD, England. *Bulletin de la Société Royale Belge d'Anthropologie et de Préhistoire* 99:53–59.

Molleson T. and Cox M. (1993) *The Spitalfields Project*, vol. 2, *The Middling Sort*, Research Report no. 86. York, UK: Council for British Archaeology.

Molleson T. and Jones K. (1991) Dental evidence for dietary change at Abu Hureyra. *Journal of Archaeological Science* 18:525–539.

Molleson T., Cruse K. and Mays S. (1998) Some sexually dimorphic features of the human juvenile skull and their value in sex determination in immature skeletal remains. *Journal of Archaeological Science* 25:719–728.

Molto J. (2000) Humerus varus deformity in Roman period burials from Kellis 2, Dakhleh, Egypt. *American Journal of Physical Anthropology* 113:103–109.

Montgomery J., Budd P. and Evans J. (2000) Reconstructing the lifetime movements of ancient people: a Neolithic case from Southern England. *European Journal of Archaeology* 3:370–385.

Moon H. (1877) On irregular and defective tooth development. *Transactions of the Odontological Society* **9**:223–243.

Moore J. and Scott E. (eds.) (1997) *Invisible People and Processes.* Leicester, UK: Leicester University Press.

Moore J. A., Swedlund A. C. and Armelagos G. J. (1975) The use of life tables in paleodemography. *American Antiquity* **39**:57–70.

Moorrees C. F. A., Fanning E. A. and Hunt E. E. (1963a) Formation and resorption of three deciduous teeth in children. *American Journal of Physical Anthropology* **21**:205–213.

Moorrees C. F. A., Fanning E. A. and Hunt E. E. (1963b) Age variation of formation stages for ten permanent teeth. *Journal of Dental Research* **42**:1490–1502.

Morton R. (1689) *Phthisiologica.* London.

Morton R. J. and Lord W. (2002) Detection and recovery of abducted and murdered children: behavioral and taphonomic influences. In W. Haglund and M. Sorg (eds.) *Advances in Forensic Taphonomy: Method, Theory and Archaeological Perspectives.* New York: CRC Press, pp. 151–171.

Moseley J. E. (1974) Skeletal changes in the anemias. *Seminars in Roentgenology* **9**:169–184.

Moss M. and Moss-Salentijn L. (1977) Analysis of developmental processes possibly related to human sexual dimorphism in permanent and deciduous canines. *American Journal of Physical Anthropology* **46**:407–414.

Mulder E., Robles de Medina P., Huizink A., *et al.* (2002) Prenatal maternal stress: effects on pregnancy and the (unborn) child. *Early Human Development* **70**:3–14.

Mulhern D. (2002) Probable case of Binder Syndrome in a skeleton from Quarai, New Mexico. *American Journal of Physical Anthropology* **118**:371–377.

Mulinski T. M. J. (1976) The use of fetal material as a measure of stress at Grasshopper Pueblo. Paper presented at the *Society for American Archaeology*, St Louis, IL.

Muller-Bolla M., Lupi-Pegurier L., Quatrehomme G., Velly A. and Bolla M. (2003) Age estimation from teeth in children and adolescents. *Journal of Forensic Sciences* **48**:4–13.

Murail P. and Girard L. (2000) Biology and burial practices from the end of the 1st century AD to the beginning of the 5th century AD: the rural cemetery of Chantambre (Essonne, France). In J. Pearce, M. Millett and M. Struck (eds.) *Burial, Society and Context in the Roman World.* Oxford, UK: Oxbow Books, pp. 105–111.

Narchi H., El Jamil M. and Kulaylat N. (2001) Symptomatic rickets in adolescence. *Archives of Diseases in Childhood* **84**:501–503.

Nathan H. and Haas N. (1966) On the presence of cribra orbitalia in apes and monkeys. *American Journal of Physical Anthropology* **24**:351–360.

Nava A., Bondioli L., Fitzgerald C. M. and Macchiarelli R. (2005) Childhood health in the community of Porteus Romae (2nd to 3rd century BCE) determined from microscopic defects in children with mixed dentitions. *American Journal of Physical Anthropology* Suppl. **40**:155–156.

Nawrocki S. P. (1995) Taphonomic processes in historic cemeteries. In A. L. Grauer (ed.) *Bodies of Evidence: Reconstructing History through Skeletal Analysis.* New York: John Wiley, pp. 49–66.

Nerlich A. and Zink A. (1995) Evidence of Langerhans cell histocytosis in an infant of a late Roman cemetery. *Journal of Paleopathology* **7**:119.

Neuhauser E. (1970) Infantile cortical hyperostosis and skull defects. *Postgraduate Medicine* **48**:57–59.

Newman D. (2004) *Sociology: Exploring the Architecture of Everyday Life*. New York: Pine Forge Press.

Newman J. (1995) How breastmilk protects newborns. *Scientific American* **273**:76–80.

Nicholls T. (2003) A re-examination of sex determination methods for juvenile skeletons using archaeological samples. M.Sc. dissertation, Bournemouth University, Bournemouth, UK.

Nichols R., Townsend E. and Malina R. (1983) Development of permanent teeth in Mexican–American children. *American Journal of Physical Anthropology* Suppl. **60**:232.

Nielsen-Marsh C., Gernaey A., Turner-Walker G., *et al.* (2000) The chemical degradation of bone. In M. Cox and S. Mays (eds.) *Human Osteology in Archaeology and Forensic Science*. London: Greenwich Medical Media Ltd, pp. 439–454.

Niswander J. D. (1965) Permanent tooth eruption in children with major physical defect and disease. *Journal of Dentistry for Children* **32**:266–268.

Nolla C. (1960) The development of the permanent teeth. *Journal of Dentistry for Children* **27**:254–256.

Noren J. (1983) Enamel structure in deciduous teeth from low-birth-weight infants. *Acta Odontologica Scandinavia* **41**:355–362.

Noren J., Magnusson B. and Grahnen H. (1978) Mineralisation defects in primary teeth in intra-uterine undernutrition. II. A histological and microradiographic study. *Swedish Dental Journal* **2**:67–72.

Norman N. (2002) Death and burial of Roman children: the case of the Yasima cemetery at Carthage. I. Setting the stage. *Mortality* **7**:303–323.

Norman N. (2003) Death and burial of Roman children: the case of the Yasima cemetery at Carthage. II. The archaeological evidence. *Mortality* **8**:36–47.

Norris S. (2002) Mandibular ramus height as an indicator of human infant age. *Journal of Forensic Sciences* **47**:8–11.

Nowak O. and Piontek J. (2002) The frequency of appearance of transverse (Harris) lines in the tibia in relationship to age at death. *Annals of Human Biology* **29**:314–325.

Núñez-de la Mora A., Napolitano D., Choudary O. and Bentley G. (2005) Changes in breastfeeding practices among migrant Bangladeshi women in London. *American Journal of Physical Anthropology* Suppl. **40**:159.

Nykänen R., Espeland L., Kvaal S. and Krogstad O. (1998) Validity of the Demirjian methods for dental age estimation when applied to Norwegian children. *Acta Odontologica Scandinavia* **56**:238–244.

Nyström M., Peck L., Kleemola-Kujala E., Evälahti M. and Kataja M. (2000) Age estimation in small children: reference values based on counts of deciduous teeth in Finns. *Forensic Science International* **110**:179–188.

Nyström M. and Ranta H. (2003) Tooth formation and the mandibular symphysis during the first five postnatal months. *Journal of Forensic Sciences* **48**:1–5.

O'Brien M. (1994) *Children's Dental Health in the United Kingdom 1993*. London: Her Majesty's Stationery Office.

O'Connell T. and Hedges R. (1999) Investigations into the effect of diet on modern human hair isotopic values. *American Journal of Physical Anthropology* **106**:409–426.

O'Connor J. and Cohen J. (1987) Dating fractures. In P. Kleinman (ed.) *Diagnostic Imaging of Child Abuse*. Baltimore, MD: Williams & Wilkins, pp. 103–113.

Ogden J. (1981) Injury to the growth mechanism of the immature skeleton. *Skeletal Radiology* **6**:237.

Olivier G. (1969) *Practical Anthropology*. Springfield, IL: Charles C. Thomas.

Olivier G. and Pineau H. (1960) Nouvelle determination de la taille foetale d'après les longueurs diaphysaires des os longs. *Annales die Médecine Légale* **40**:141–144.

Orme N. (2001) *Medieval Children*. New Haven, CT: Yale University Press.

Ortner D. J. (1984) Bone lesions in a probable case of scurvy from Metlatavik, Alaska. *Museum of Applied Science Center for Archaeology Journal* **3**:79–81.

Ortner D. J. (1991) Theoretical and methodological issues in paleopathology. In D. J. Ortner and A. C. Aufderheide (eds.) *Human Paleopathology: Current Syntheses and Future Options*. Washington, DC: Smithsonian Institution Press, pp. 5–11.

Ortner D. J. (2003) *Identification of Pathological Conditions in Human Skeletal Remains*. New York: Academic Press.

Ortner D. J. and Ericksen M. F. (1997) Bone changes in the human skull probably resulting from scurvy in infancy and childhood. *International Journal of Osteoarchaeology* **7**:212–220.

Ortner D. J. and Mays S. (1998) Dry-bone manifestations of rickets in infancy and early childhood. *International Journal of Osteoarchaeology* **8**:45–55.

Ortner D. J. and Putschar W. G. J. (1985) *Identification of Pathological Conditions in Human Skeletal Remains*. Washington, DC: Smithsonian Institution Press.

Ortner D., Kimmerle E. and Diez M. (1999) Probable evidence of scurvy in subadults from archaeological sites in Peru. *American Journal of Physical Anthropology* **108**:321–331.

Ortner D. J., Butler W., Cafarella J. and Milligan L. (2001) Evidence of probable scurvy in subadults from archaeological sites in North America. *American Journal of Physical Anthropology* **114**:343–351.

O'Sullivan E. A., Williams S. A. and Curzon M. E. J. (1989) Dental caries and nutritional stress in English archaeological child populations. In C. A. Roberts, F. Lee and J. Bintliff (eds.) *Burial Archaeology: Current Research, Methods and Developments*, BAR (British Series) no. 211. Oxford: Archaeopress, pp. 167–174.

O'Sullivan E. A., Williams S. A. and Curzon M. E. J. (1992) Dental caries in relation to nutritional stress in early English child populations. *Pediatric Dentistry* **14**:26–29.

Owsley D. and Bass W. (1979) A demographic analysis of skeletons from the Larson site (39WW2) Walworth County, South Dakota: vital statistics. *American Journal of Physical Anthropology* **51**:145–154.

Owsley D. W. and Bradtmiller B. (1983) Mortality of pregnant females in Arikara villages: osteological evidence. *American Journal of Physical Anthropology* **61**:331–336.

Owsley D. and Jantz R. (1983) Formation of the permanent dentition in Arikara Indians: timing differences that affect dental age assessments. *American Journal of Physical Anthropology* **61**:467–471.

Owsley D. W. and Jantz R. L. (1985) Long bone lengths and gestational age distributions of post-contact period Arikara perinatal infant skeletons. *American Journal of Physical Anthropology* **68**:321–328.

Owsley D., Ubelaker D., Houck M., *et al.* (1995) The role of forensic anthropology in the recovery and analysis of Branch Davidian compound victims: techniques of analysis. *Journal of Forensic Sciences* **40**:341–348.

Özbek M. (2001) Cranial deformation in a subadult sample from Degirmentepe (Chalcolithic, Turkey). *American Journal of Physical Anthropology* **115**:238–244.

Ozden H., Balci Y., Demirüstü C., Turgut A. and Ertugrul M. (2005) Stature and sex estimates using foot and shoe dimensions. *Forensic Science International* **147**:181–184.

Ozonoff B. (1979) *Pediatric Orthopedic Radiology.* Philadelphia, PA: W. B. Saunders.

Paine R. (2000) If a population crashes in prehistory, and there is no palaeodemographer there to hear it, does it make a sound? *American Journal of Physical Anthropology* **112**:181–190.

Pálfi G., Ardagna Y., Maczel Y., *et al.* (2000) Traces des infections osseuses dans la série anthropologique de la Celle (Var, France): résultats preliminaries. Paper presented at the *Colloque 2000 du Groupe des Paléopathologistes de Langue Française,* February 11–13.

Pálfi G., Dutour O., Borreani M., Brun J.-P. and Berato J. (1992) Pre-Columbian congenital syphilis from the Late Antiquity in France. *International Journal of Osteoarchaeology* **2**:245–261.

Palkama A., Hopsu V., Takki S. and Takki K. (1965) Children's age and stature estimated from femur diameter. *Annales de Médécine Experimentalis et Biologiae Fenniae* **44**:86–87.

Palkovich A. (1980) *Pueblo Population and Society: Arroyo Hondo Skeletal and Mortuary Remains,* Arroyo Hondo Archaeological Series no. 3. Santa Fe, NM: School of American Research Press.

Palkovich A. (1987) Endemic disease patterns in palaeopathology: porotic hyperostosis. *American Journal of Physical Anthropology* **74**:527–537.

Palm T. A. (1890) The geographical distribution and aetiology of rickets. *Practitioner* **45**:270–321.

Palmirotta R., Verginellii F., Tota G., *et al.* (1997) Use of a multiple polymerase chain reaction assay in the sex typing of DNA extracted from archaeological bone. *International Journal of Osteoarchaeology* **7**:605–609.

Palubeckaité Z., Jankaukas R. and Boldsen J. (2002) Enamel hypoplasia in Danish and Lithuanian late medieval/early modern samples: a possible reflection of child morbidity and mortality patterns. *International Journal of Osteoarchaeology* **12**:189–201.

Panhuysen R. (1999) Child mortality in early medieval Maastricht: missing children? *Journal of Paleopathology* **11**:94.

Panhuysen R., Coenen V. and Bruintjes T. D. (1997) Chronic maxillary sinusitis in medieval Maastricht, The Netherlands. *International Journal of Osteoarchaeology* **7**:610–614.

Panter-Brick C. (1997) Seasonal growth patterns in rural Nepali children. *Annals of Human Biology* **24**:1–18.

Panter-Brick C., Lunn P., Baker R. and Todd A. (2001) Elevated acute-phase protein in stunted Nepali children reporting low morbidity: different rural and urban profiles. *British Journal of Nutrition* **85**:125–131.

Panter-Brick C., Todd A. and Baker R. (1996) Growth status of homeless Nepali boys: do they differ from rural and urban controls? *Social Science of Medicine* **43**:441–451.

Park E. A. (1954) Bone growth in health and disease. *Archives of Diseases in Childhood* **29**:269–281.

Park E. A. (1964) The imprinting of nutritional disturbances on the growing bone. *Pediatrics* **33**:815–862.

Park E. A. and Richter C. P. (1953) Transverse lines in bone: the mechanism of their development. *Bulletin of the Johns Hopkins Hospital* **93**:234–248.

Parry J. S. (1872) Remarks on the pathological anatomy, causes, and treatment of rickets. *American Journal of Medical Sciences* **126**:305–329.

Paterson R. (1929) A radiological investigation of the epiphyses of the long bones. *Journal of Anatomy* **64**:28–46.

Penn W. (1960) Springhead: Temples III and IV. *Archaeologia Cantiana* **74**:113–140.

Penny M. E. and Lanata C. F. (1995) Zinc in the management of diarrhea in young children. *New England Journal of Medicine* **333**:873–874.

Perzigian A. and Jolly P. (1984) Skeletal and dental identification of an adolescent female. In T. A. Rathburn and J. Buikstra (eds.) *Human Identification: Case Studies in Forensic Anthropology.* Springfield, IL: Charles C. Thomas, pp. 244–252.

Pettifor J. (2003) Nutritional rickets. In F. Glorieux, J. Pettifor and M. Jüppner (eds.) *Pediatric Bone: Biology and Diseases.* New York: Academic Press, pp. 541–565.

Pettifor J. and Daniels E. (1997) Vitamin D deficiency and nutritional rickets in children. In D. Feldman, F. Glorieux and J. Pike (eds.) *Vitamin D.* New York: Academic Press, pp. 663–678.

Pfeiffer S. (1991) Rib lesions and New World tuberculosis. *International Journal of Osteoarchaeology* **1**:191–198.

Pfeiffer S. and Crowder C. (2004) An ill child among mid-Holocene foragers of Southern Africa. *American Journal of Physical Anthropology* **123**:23–29.

Piercecchi-Marti M.-D., Adalian P., Bourliere-Najean B., *et al.* (2002) Validation of a radiographic method to establish new fetal growth standards: radio-anatomical correlation. *Journal of Forensic Sciences* **47**:328–331.

Piontek J. and Kozlowski T. (2002) Frequency of cribra orbitalia in the subadult medieval population from Gruczno, Poland. *International Journal of Osteoarchaeology* **12**:202–208.

Pitt M. J. (1988) Rickets and osteomalacia. In D. Resnick and G. Niwayama (eds.) *Diagnosis of Bone and Joint Disorders.* Philadelphia, PA: W. B. Saunders, pp. 2087–2126.

Plunkett J. and Plunkett M. (2000) Physiologic periosteal changes in infancy. *American Journal of Forensic Pathology* **21**:213–216.

Ponec D. J. and Resnick D. (1984) On the etiology and pathogenesis of porotic hyperostosis of the skull. *Investigative Radiology* **19**:313–317.

Post J. B. (1971) Ages at menarche and menopause: some medieval authorities. *Population Studies* **25**:83–87.

Powell F. (1996) The human remains. In A. Boddington (ed.) *Raunds Furnells: The Anglo-Saxon Church and Churchyard.* London: English Heritage, pp. 113–124.

Powell M. L. and Cook D. (2005) Treponematosis: inquires into the nature of a protean disease. In M. Powell and D. Cook (eds.) *The Myth of Syphilis: The Natural History of Treponematosis in North America.* Gainesville, FL: University of Florida Press, pp. 9–62.

Powers R. (1980) A tool for coping with juvenile human bones from archaeological excavations. In W. White (ed.) *Skeletal Remains from the Cemetery of St. Nicolas Shambles, City of London.* London: London and Middlesex Archaeological Trust, pp. 74–78.

Prader A. and Budlinger H. (1977) Body measurements, growth velocity and bone age of healthy children up to 12 years of age (longitudinal growth study, Zurich). *Helvetica Paediatrica Acta* **37**:5–44.

Prader A., Tanner J. M. and von Harnack G. A. (1963) Catch-up growth following illness or starvation. *Journal of Pediatrics* **62**:646–659.

Prentice A. (2003) Pregnancy and lactation. In F. Glorieux, J. Pettifor and M. Jüppner (eds.) *Pediatric Bone: Biology and Diseases.* New York: Academic Press, pp. 249–269.

Quant S. A. (1996) Nutrition in medical anthropology. In C. F. Sargent and T. M. Johnson (eds.) *Medical Anthropology: Contemporary Theory and Method.* London: Praeger, pp. 272–289.

Radosevich S. (1993) The six deadly sins of trace element analysis: a case of wishful thinking in science. In M. Sandford (ed.) *Investigations of Ancient Human Tissue: Chemical Analysis in Anthropology.* Langthorne, UK: Gordon and Breach, pp. 269–332.

Rallison M. L. (1986) *Growth Disorders in Infants, Children and Adolescents.* Salt Lake City, UT: John Wiley.

Rasool M. (2001) Primary subacute haematogenous osteomyelitis in children. *Journal of Bone and Joint Surgery* **83**-B:93–98.

Redfield A. (1970) A new aid to aging immature skeletons: development of the occipital bone. *American Journal of Physical Anthropology* **33**:207–220.

Rega E. (1997) Age, gender and biological reality in the early Bronze Age cemetery at Morkin. In J. Moore and E. Scott (eds.) *Invisible People and Processes.* Leicester, UK: Leicester University Press, pp. 229–247.

Reid D. and Dean M. (2000) The timing of linear hypoplasias on human anterior teeth. *American Journal of Physical Anthropology* **113**:135–141.

Reinhard K. J. (1992) Patterns of diet, parasitism and anemia in prehistoric West North America. In P. Stuart-Macadam and S. Kent (eds.) *Diet, Demography and Disease: Changing Patterns of Anemia.* New York: Aldine de Gruyter, pp. 219–260.

Reinken H., Stolley H., Droese W. and van Oost G. (1980) Longitudinale Körperentwicklung gesunder Kinder. II. Grösse, gwicht, hautfettfalter von kindern in alter von 1, 5 bis 16 jahren. *Klinische Pädiatrie* **192**:25–33.

Resnick D. and Goergen T. (2002) Physical injury: concepts and terminology. In D. Resnick (ed.) *Diagnosis of Bone and Joint Disorders.* Philadelphia, PA: W. B. Saunders, pp. 2627–2789.

Resnick D. and Niwayama D. (eds.) (1988) *Diagnosis of Bone and Joint Disorders.* Philadelphia, PA: W. B. Saunders.

Reynolds E. (1945) The bony pelvic girdle in early infancy. *American Journal of Physical Anthropology* **3**:321–354.

Reynolds E. (1947) The bony pelvis in prepubertal children. *American Journal of Physical Anthropology* **5**:165–200.

Ribot I. and Roberts C. A. (1996) A study of non-specific stress indicators and skeletal growth in two mediaeval subadult populations. *Journal of Archaeological Science* **23**:67–79.

Richards M. (2001) How distinctive is genetic information? *Studies in History and Philosophy of Biological and Biomedical Sciences* **32**:663–687.

Richards M., Mays S. and Fuller B. (2002) Stable carbon and nitrogen isotope values of bone and teeth reflect weaning age at the medieval Wharram Percy site, Yorkshire, UK. *American Journal of Physical Anthropology* **119**:205–210.

Rikhasor R., Qureshi A., Rathi S. and Channa N. (1999) Skeletal maturity in Pakistani children. *Journal of Anatomy* **195**:305–308.

Rissech C., Garcia M. and Malgosa A. (2003) Sex and age diagnosis by ischium morphometric analysis. *Forensic Science International* **135**:188–196.

Robbins L. (1977) The story of life revealed by the dead. In R. Blakey (ed.) *Biocultural Adaptation in Prehistoric America.* Athens, GA: University of Georgia Press, pp. 10–26.

Roberts C. (1987) Case Report No. 9. *Paleopathology Newsletter* **57**:14–15.

Roberts C. (1989) The human remains from 76 Kingsholm, Gloucester, unpublished skeletal report. Bradford, UK: University of Bradford.

Roberts C. (2000) Trauma in biocultural perspective: past present and future work in Britain. In M. Cox and S. Mays (eds.) *Human Osteology in Archaeology and Forensic Science.* London: Greenwich Medical Media Ltd, pp. 337–356.

Roberts C. and Buikstra J. (2003) *The Bioarchaeology of Tuberculosis: A Global View on a Reemerging Disease.* Gainesville, FL: University of Florida Press.

Roberts C. and Cox M. (2003) *Health and Disease in Britain.* Stroud, UK: Alan Sutton Publishing.

Roberts C. and Manchester K. (2005) *The Archaeology of Disease*, 3rd edn. Stroud, UK: Alan Sutton Publishing.

Roberts C., Boylston A., Buckley L., Chamberlain A. C. and Murphy E. M. (1998) Rib lesions and tuberculosis: the palaeopathological evidence. *Tubercle and Lung Disease* **79**:55–60.

Robinson S., Nicholson R., Pollard A. and O'Connor T. (2003) An evaluation of nitrogen porosimetry as a technique for predicting taphonomic durability in animal bone. *Journal of Archaeological Science* **30**:391–403.

Robson S. (2004) Breast milk, diet and large human brains. *Current Anthropology* **45**:419–425.

Roche A. F. (1978) Bone growth and maturation. In F. Falkner and J. M. Tanner (eds.) *Human Growth: A Comprehensive Treatise.* New York: Plenum Press, pp. 317–355.

Roede M. and van Wieringen, J. (1985) Growth diagrams, 1980. *Tijdskrift voor Sociale Gezondheidszorg* **63**:1–34.

Roelants M. and Hauspie R. (2004) *Flemish Growth Charts 2–20 Years: Use and Interpretation*. Brussels: Laboratorium voor Antropogenetica, University of Brussels.

Rogers J. (1984) Skeletons from the lay cemetery at Taunton Priory. In P. Leach (ed.) *The Archaeology of Taunton: Excavations and Fieldwork to 1980*. Taunton, UK: Western Archaeological Trust, pp. 100–120.

Rogers J. (1999) Burials: the human skeletons. In C. Heighway and R. Bryant (eds.) *The Golden Minster: The Anglo-Saxon Minster and Later Medieval Priory of St Oswald, Gloucester*. York, UK: Council for British Archaeology, pp. 229–246.

Rogers J. and Waldron T. (1988) Two possible cases of infantile cortical hyperostosis. *Paleopathology Newsletter* **63**:9–12.

Rona R. and Altman D. (1977) National study of health and growth: standards of attained height, weight and triceps skinfold in English children, 5 to 11 years old. *Annals of Human Biology* **4**:501–524.

Ronalds G., Phillips D., Godfrey K. and Manning J. (2002) The ratio of second to fourth digit lengths: a marker of impaired fetal growth. *Early Human Growth and Development* **68**:21–26.

Ronsmans C., Lewis G., Hurt L., *et al*. (2002) Mortality in pregnant and nonpregnant women in England and Wales 1997–2002: are pregnant women healthier? In Confidential Enquiry into Maternal and Child Health (CEMACH) (ed.) *Why Mothers Die 2000–2002: Report on Confidential Enquires into Maternal Deaths in the United Kingdom*. London: Royal College of Obstetricians and Gynaecologists, pp. 120–126.

Rose J. C., Armelagos G. J. and Lallo J. W. (1978) Histological enamel indicator of childhood stress in prehistoric skeletal samples. *American Journal of Physical Anthropology* **49**:511–519.

Rose J. C., Condon K. W. and Goodman A. H. (1985) Diet and dentition: developmental disturbances. In R. I. Gilbert and J. M. Meilke (eds.) *The Analysis of Prehistoric Diets*. Orlando, FL: Academic Press, pp. 281–305.

Rosenzweig K. and Garbarski D. (1965) Numerical aberrations in the permanent teeth of grade school children in Jerusalem. *American Journal of Physical Anthropology* **23**:277–284.

Roske G. (1930) Eine eigenartige Knochenerkrankung im Säuglingsalter. *Monatsschrift Kinderheilkunde* **47**:385–393.

Rossi F. (1993) Deux poupées en ivoire d'époque romaine à Yverdon-les-Bains VD. *Archäologie der Schweiz* **16**:152–157.

Roth E. A. (1992) Applications of demography models to paleodemography. In S. R. Saunders and M. A. Katzenberg (eds.) *Skeletal Biology of Past Peoples: Research Methods*. New York: Wiley-Liss, pp. 175–188.

Rothschild B., Hershkovitz I., Bedford L., *et al*. (1997) Identification of childhood arthritis in archaeological material: juvenile rheumatoid arthritis versus juvenile spondyloarthropathy. *American Journal of Physical Anthropology* **102**:249–264.

Rothschild B. M. and Rothschild C. (1997) Congenital syphilis in the archaeological record: diagnostic insensitivity of osseous lesions. *International Journal of Osteoarchaeology* **7**:39–42.

Roveland B. (2000) Footprints in the clay: Upper Palaeolithic children in ritual and secular contexts. In J. Sofaer Derevenski (ed.) *Children and Material Culture*. London: Routledge, pp. 29–38.

Rowe P. (2001) Why is rickets resurgent in the USA? *Lancet* **357**:1100.

Roy T., Ruff C. and Palto C. (1994) Hand dominance and bilateral asymmetry in the structure of the second metacarpal. *American Journal of Physical Anthropology* **94**:203–211.

Rudney J. D. (1983) Dental indicators of growth disturbance in a series of ancient Lower Nubian populations: changes over time. *American Journal of Physical Anthropology* **60**:463–470.

Ruff C. (2000) Biomechanical analyses of archaeological human skeletons. In A. Katzenberg and S. R. Saunders (eds.) *Biological Anthropology of the Human Skeleton*. New York: Wiley-Liss, pp. 71–102.

Rushton J. (1991) The secret 'iron tongs' of midwifery. *Historian* **30**:12–15.

Rushton M. A. (1933) Fine contour-lines of enamel in milk teeth. *Dental Record* **53**:170–171.

Rushton M. A. (1965) The teeth of Anne Mowbray. *British Dental Journal* **119**:355–359.

Rutty G. (2001) Postmortem changes and artefacts. In G. Rutty (ed.) *Essentials of Autopsy Practice*. London: Springer-Verlag, p. 66.

Ryan A. S. (1997) Iron-deficiency anemia in infant development: implications for growth, cognitive development, resistance to infection, and iron supplementation. *Yearbook of Physical Anthropology* **40**:25–62.

Ryan W. (1862) *Infanticide: Its Law, Prevalence, Prevention, and History*. London.

Saarinen U. M. (1978) Need for iron supplementation in infants on prolonged breast feeding. *Journal of Pediatrics* **93**:177–180.

Salter R. (1980) Special features of fractures and dislocation in children. In R. Heppenstall (ed.) *Fracture Treatment and Healing*. Philadelphia, PA: W. B. Saunders, p. 190.

Salter R. and Harris W. (1963) Injuries involving the epiphyseal plate. *Journal of Bone and Joint Surgery* **45**:587–622.

Sandin-Dominguez M. (1988) *Curvas Semilongitudinale de Crecimiento: Ninos entre 6 y 15 Anos*. Madrid: Universidad Autónoma de Madrid.

Santos A. L. and Roberts C. A. (2001) A picture of tuberculosis in young Portuguese people in the early 20th century: a multidisciplinary study of the skeletal and historical evidence. *American Journal of Physical Anthropology* **115**:38–49.

Santos R. V. and Coimbra C. E. A. (1999) Hardships of contact: enamel hypoplasias in Tupí-Mondé Amerindians from the Brazilian Amazonia. *American Journal of Physical Anthropology* **109**:111–127.

Sarnat B. G. and Schour I. (1941) Enamel hypoplasia (chronological enamel aplasia) in relation to systemic disease: a chronologic, morphologic and etiologic classification. *Journal of the American Dental Association* **28**:1989–2000.

Saternus K.-S., Kernbach-Wighton G. and Oehmichen M. (2000) The shaking trauma in infants: kinetic chains. *Forensic Science International* **109**:203–213.

Sauer R. (1978) Infanticide and abortion in nineteenth-century Britain. *Population Studies* **32**:81–93.

Saunders S. R. (1992) Subadult skeletons and growth related studies. In S. R. Saunders and M. A. Katzenberg (eds.) *Skeletal Biology of Past Populations: Advances in Research Methods*. New York: Wiley-Liss, pp. 1–20.

Saunders S. R. (2000) Subadult skeletons and growth-related studies. In M. A. Katzenberg and S. R. Saunders (eds.) *Biological Anthropology of the Human Skeleton.* New York: Wiley-Liss, pp. 1–20.

Saunders S. R. and Barrans L. (1999) What can be done about the infant category in skeletal samples? In R. D. Hoppa and C. M. Fitzgerald (eds.) *Human Growth in the Past: Studies from Bones and Teeth.* Cambridge, UK: Cambridge University Press, pp. 183–209.

Saunders S. R. and Hoppa R. D. (1993) Growth deficit in survivors and non-survivors: biological mortality bias in subadult skeletal samples. *Yearbook of Physical Anthropology* **36**:127–151.

Saunders S. R. and Melbye F. J. (1990) Subadult mortality and skeletal indicators of health in Late Woodland Ontario Iroquois. *Canadian Journal of Archaeology* **14**:61–74.

Saunders S. R. and Yang D. (1999) Sex determination: XX or XY from the human skeleton. In S. I. Fairgrieve (ed.) *Forensic Osteological Analysis: A Book of Case Studies.* Springfield, IL: Charles C. Thomas, pp. 36–59.

Saunders S. R., Hoppa R. and Southern R. (1993a) Diaphyseal growth in a nineteenth century skeletal sample of subadults from St Thomas' Church, Belleville. *International Journal of Osteoarchaeology* **3**:265–281.

Saunders S. R., DeVito C., Herring A. D., Southern R. and Hoppa R. D. (1993b) Accuracy tests of tooth formation age estimations for human skeletal remains. *American Journal of Physical Anthropology* **92**:173–188.

Saunders S. R., Herring D. A. and Boyce G. (1995) Can skeletal samples accurately represent the living populations they come from? The St. Thomas' cemetery site, Belleville, Ontario. In A. L. Grauer (ed.) *Bodies of Evidence: Reconstructing History through Skeletal Analysis.* New York: Wiley-Liss, pp. 69–89.

Sazawal S., Black R. E., Bhan M. K., *et al.* (1995) Zinc supplementation in young children with acute diarrhea in India. *New England Journal of Medicine* **333**:839–844.

Scammon R. and Calkins L. (1923) New empirical formulae for expressing the linear growth of the human fetus. *Anatomical Record* **25**:148–149.

Schamall D., Teschler-Nicola M., Kainberger F., *et al.* (2003) Changes in trabecular bone structure in rickets and osteomalacia: the potential of a medico-historical collection. *International Journal of Osteoarchaeology* **13**:283–288.

Schell L. M. (1981) Environmental noise and human prenatal growth. *American Journal of Physical Anthropology* **56**:63–70.

Schell L. M. (1998) Urbanisation and growth. In S. Ulijaszek, F. E. Johnston and M. A. Preece (eds.) *The Cambridge Encyclopedia of Human Growth and Development.* Cambridge,UK: Cambridge University Press, pp. 408–409.

Scheuer L. (1998) Age at death and cause of death of the people buried in St Bride's Church, Fleet Street, London. In M. Cox (ed.) *Grave Concerns: Death and Burial in England 1700 to 1850.* York, UK: Council for British Archaeology, pp. 100–111.

Scheuer L. (2002) Brief communication: a blind test of mandibular morphology for sexing mandibles in the first few years of life. *American Journal of Physical Anthropology* **119**:189–191.

Scheuer L. and Black S. (2000) *Developmental Juvenile Osteology*. London: Academic Press.

Scheuer L. and Black S. (2004) *The Juvenile Skeleton*. London: Elsevier.

Scheuer L. and Maclaughlin-Black S. (1994) Age estimation from the pars basilaris of the fetal and juvenile occipital bone. *International Journal of Osteoarchaeology* 4:377–380.

Scheuer L., Musgrave J. H. and Evans S. P. (1980) The estimation of late fetal and perinatal age from limb bone length by linear and logarithmic regression. *Annals of Human Biology* 7:257–265.

Schmeling A., Olze A., Reisinger W. and Geserick G. (2001) Age estimation of living people undergoing criminal proceedings. *Lancet* 358:89–90.

Schmeling A., Reisinger W., Loreck D., *et al.* (2000) Effects of ethnicity on skeletal maturation: consequences for forensic age estimations. *International Journal of Legal Medicine* 113:253–258.

Schmidt E. (1892) Die Körpergrösse und das Gewicht der Schulkinder des kreises Saalfeld. *Archiv für Anthropologie* 21:385–434.

Schmidt P., Müller R., Dettmeyer R. and Madea B. (2002) Suicide in children, adolescents and young adults. *Forensic Science International* 127:161–167.

Schmidt-Schultz T. and Schultz M. (2003) Bone protects proteins over thousands of years: extraction, analysis and interpretation of extracellular matrix proteins in archaeological skeletal remains. *American Journal of Physical Anthropology* 123:30–39.

Schmitt P. and Glorion C. (2004) Osteomyelitis in infants and children. *European Radiology* 14:44–54.

Schmitt S. (1998) *Statistics of a Mass Grave*. Available online at http://garnet.acns.fsu.edu/ ~sss4407/Rwanda/RWStats.htm/

Schofield R. and Wrigley E. (1979) Infant and child mortality in England in the late Tudor and early Stuart period. In C. Webster (ed.) *Health, Medicine and Mortality in the Sixteenth Century*. Cambridge, UK: Cambridge University Press, pp. 61–95.

Schönau E. and Rauch F. (2003) Biochemical markers of bone metabolism. In F. Glorieux, J. Pettifor and M. Jüppner (eds.) *Pediatric Bone: Biology and Diseases*. New York: Academic Press, pp. 339–357.

Schour I. and Massler M. (1941) The development of the human dentition. *Journal of the American Dental Association* 28:1153–1160.

Schulman I. (1959) The anemia of prematurity. *Journal of Pediatrics* 54:663–672.

Schultz A. (1923) Fetal growth in man. *American Journal of Physical Anthropology* 6:389–399.

Schultz M. (1984) The diseases in a series of children's skeletons from Ikiz Tepe, Turkey. In V. Capecchi and E. Rabino Massa (eds.) *Proceedings of the 5th European Meeting of the Paleopathology Association, Siena, Italy*. Siena: Tipografia Sienese, pp. 321–325.

Schultz M. (1989) Causes and frequency of diseases during early childhood in Bronze Age populations. In L. Capasso (ed.) *Advances in Palaeopathology*. Chieti, Italy: Marino Solfanelli Editore, pp. 175–179.

Schultz M. (1993a) Initial stages of systemic bone disease. In G. Grupe and A. N. Garland (eds.) *Histology of Ancient Human Bone: Methods and Diagnosis*. New York: Springer-Verlag, pp. 185–203.

Schultz M. (1993b) *Vestiges of Non-Specific Inflammation in Prehistoric and Historic Skulls: A Contribution to Palaeopathology*. Aesch, Switzerland: Anthropologische Beitrage.

Schultz M. (1994) Comparative histopathology of syphilitic lesions in prehistoric and historic human bones. In O. Dutour and G. Palfi (eds.) *The Origin of Syphilis in Europe: Before or After 1493?* Toulon, France: Université de Provence, pp. 63–67.

Schultz M. (1997) Porotic hyperostosis in Spanish Florida: nature and etiology of a frequently observed phenomenon. *American Journal of Physical Anthropology* Suppl. **24**:206.

Schultz M. (2001) Palaeohistopathology of bone: a new approach to the study of ancient diseases. *Yearbook of Physical Anthropology* **44**:106–147.

Schurr M. (1997) Stable nitrogen isotopes as evidence for the age of weaning at the Angel Site: a comparison of isotopic and demographic measures of weaning age. *Journal of Archaeological Science* **24**:919–927.

Schurr M. (1998) Using stable nitrogen isotopes to study weaning behaviour in past populations. *World Archaeology* **30**:327–342.

Schurr M. and Powell M. (2005) The role of changing childhood diets in the prehistoric evolution of food production: an isotopic assessment. *American Journal of Physical Anthropology* **126**:278–294.

Schutkowski H. (1993) Sex determination of infant and juvenile skeletons. I. Morphological features. *American Journal of Physical Anthropology* **90**:199–205.

Sciulli P. (1994) Standardization of long bone growth in children. *International Journal of Osteoarchaeology* **4**:257–259.

Sciulli P. and Oberly J. (2002) Native Americans in Eastern North America: the Southern Great Lakes and Upper Ohio Valley. In R. H. Steckel and J. C. Rose (eds.) *The Backbone of History: Health and Nutrition in the Western Hemisphere*. Cambridge, UK: Cambridge University Press, pp. 440–480.

Scott E. (1979) Dental wear scoring technique. *American Journal of Physical Anthropology* **51**:213–218.

Scott E. (1990) A critical review of the interpretation of infant burials in Roman Britain, with particular reference to villas. *Journal of Theoretical Archaeology* **1**:30–46.

Scott E. (1991) Animal and infant burials in Romano-British villas: a revitalization movement. In P. Garwood, D. Jennings, R. Skeates and J. Toms (eds.) *Sacred and Profane*. Oxford, UK: Oxford University Committee for Archaeology, pp. 115–121.

Scott E. (1999) *The Archaeology of Infancy and Infant Death*, BAR (International Series) no. 819. Oxford, UK: Archaeopress.

Scott E. (2001) Killing the female? Archaeological narratives of infanticide. In B. Arnold and N. Wicker (eds.) *Gender and the Archaeology of Death*. Walnut Creek, CA: Altamira Press, pp. 3–21.

Scott S. and Duncan C. (1998) *Human Demography and Disease*. Cambridge, UK: Cambridge University Press.

Scott S. and Duncan C. (1999) Malnutrition, pregnancy and infant mortality: a biometric model. *Journal of Interdisciplinary History* **30**:37–60.

Scrimshaw S. (1984) Infanticide in human populations: societal and individual concern. In G. Hausfater and S. B. Hrdy (eds.) *Infanticide.* New York: Aldine, pp. 439–462.

Scrimshaw N. S. and Young V. R. (1989) Adaptation to low protein and energy intakes. *Human Organization* **48**:20–29.

Scrimshaw N. S., Taylor C. E. and Gordon J. E. (1959) Interactions of nutrition and infection. *American Journal of the Medical Sciences* **237**:367–403.

Scull C. (1997) Comment. In J. Hines (ed.) *The Anglo-Saxons from the Migration Period to the Eighth Century: An Ethnographic Perspective.* Woodbridge, UK: Boydell Press, p. 164.

Seckler D. (1982) Small but healthy? A basic hypothesis in the theory, measurement and policy of malnutrition. In P. V. Sukhatme (ed.) *Newer Concepts in Nutrition and their Implications for Policy.* Pune, India: Maharashtra Association for the Cultivation of Science Research Institute, pp. 127–137.

Seitz R. (1923) Relation of epiphyseal length to bone length. *American Journal of Physical Anthropology* **6**:37–49.

Sellen D. (2001) Comparisons of infant feeding patterns reported for nonindustrialised populations with current recommendations. *Journal of Nutrition* **131**:2707–2715.

Sellevold B. J. (1997) Children's skeletons and graves in Scandinavian archaeology. In G. De Boe and F. Verhaeghe (eds.) *Death and Burial in Medieval Europe.* Brugge, Belgium: I. A. P. Rapporten 2, pp. 15–25.

Seow K. W. (1992) Dental enamel defects in low birthweight children. *Journal of Paleopathology Monographic Publication* **2**:321–330.

Seymour J. (1996) Hungry for a new revolution: *New Scientist* March **30**:34–37.

Shahar S. (1992) *Childhood in the Middle Ages.* London: Routledge.

Shapiro D. and Richtsmeier J. (1997) Brief communication: a sample of pediatric skulls available for study. *American Journal of Physical Anthropology* **103**:415–416.

Shapiro F. (2001) *Pediatric Orthopedic Deformities: Basic Science, Diagnosis and Treatment.* New York: Academic Press.

Shaw H. and Bohrer S. (1979) The incidence of cone epiphyses and ivory epiphyses of the hand in Nigerian children. *American Journal of Physical Anthropology* **51**:155–162.

Sheldon W. (1936) Anaemia with bone changes in the skull. *Proceedings of the Royal Society of Medicine* **29**:743.

Sheldon W. (1943) *Diseases of Infancy and Childhood.* London: J. & A. Churchill.

Shell-Duncan B. and McDade T. (2002) Evaluation of low dietary iron as a nutritional adaptation to infectious disease. *American Journal of Physical Anthropology* Suppl. **34**:140.

Sherwood R., Meindl R., Robinson H. and May R. (2000) Fetal age: methods of estimation and effects of pathology. *American Journal of Physical Anthropology* **113**:305–315.

Shoesmith R. (1980) *Hereford City Excavations,* vol. 1, *Excavations at Castle Green.* Oxford, UK: Council for British Archaeology.

Shopfner C. E. (1966) Periosteal bone growth in normal infants: a preliminary report. *American Journal of Roentgenology, Radium Therapy and Nuclear Medicine* **97**:154–163.

Shorter E. (1976) *The Making of the Modern Family*. London: Collins.

Sillar B. (1994) Playing with God: cultural perceptions of children, play and miniatures in the Andes. *Archaeological Review from Cambridge* **13**:47–63.

Sillen A. and Smith P. (1984) Weaning patterns are reflected in strontium–calcium ratios of juvenile skeletons. *Journal of Archaeological Science* **11**:237–245.

Silverman F. (1953) The roentgen manifestations of unrecognised skeletal trauma in infants. *American Journal of Roentgenology* **69**:413–426.

Sjøvold T., Swedborg I. and Diener L. (1974) A pregnant woman from the Middle Ages with exostosis multiplex. *Ossa* **1**:3–23.

Skinner M. (1986) An enigmatic hypoplastic defect of the deciduous canine. *American Journal of Physical Anthropology* **69**:59–69.

Skinner M. (1997) Dental wear in immature late Pleistocene European hominines. *Journal of Archaeological Science* **24**:677–700.

Skinner M. and Anderson G. (1991) Individualization and enamel histology: a case report in forensic anthropology. *Journal of Forensic Sciences* **36**:939–948.

Skinner M. and Dupras T. (1993) Variation in birth timing and location of the neonatal line in human enamel. *Journal of Forensic Sciences* **38**:1383–1390.

Skinner M. and Goodman A. H. (1992) Anthropological uses of developmental defects of enamel. In S. R. Saunders and M. A. Katzenberg (eds.) *The Skeletal Biology of Past Peoples: Advances in Research Methods*. New York: Wiley-Liss, pp. 153–173.

Skinner M. and Newell E. (2003) Localized hypoplasia of the primary canine in Bonobos. Orangutans, and Gibbons. *American Journal of Physical Anthropology* **102**:61–72.

Šlaus M. (2000) Biocultural analysis of sex differences in mortality profiles and stress levels in the late medieval population from Nova Rača, Croatia. *American Journal of Physical Anthropology* **111**:193–209.

Smith B. (1992) Life history and the evolution of human maturation. *Evolutionary Anthropology* **1**:134–142.

Smith B. H. (1991) Standards of human tooth formation and dental age assessment. In M. A. Kelley and C. S. Larsen (eds.) *Advances in Dental Anthropology*. New York: Wiley-Liss, pp. 143–168.

Smith G. and Wood-Jones E. (1910) *The Archaeological Survey of Nubia: Report for 1907–1908*, vol. 2, *Report on the Human Remains*. Cairo: National Printing Department.

Smith P. and Avishai G. (2005) The use of dental criteria for estimating postnatal survival in skeletal remains of infants. *Journal of Archaeological Science* **32**:83–89.

Smith P. and Kahila G. (1992) Identification of infanticide in archaeological sites: a case study from the Late Roman–Early Byzantine periods at Ashkelon, Israel. *Journal of Archaeological Science* **19**:667–675.

Smith S. (1955) *Forensic Medicine*. London: J. & A. Churchill.

Smyth F., Potter A. and Silverman W. (1946) Periosteal reaction, fever and irritability in young infants: a new syndrome? *American Journal of Diseases of Children* **71**:333–350.

Snedecor S., Knapp R. and Wilson H. (1935) Traumatic ossifying periostitis of the newborn. *Surgical Gynecological Obstetrics* **61**:385–387.

Snow C. and Luke J. (1970) The Oklahoma City child disappearances of 1967: forensic anthropology in the identification of skeletal remains. In C. Rathburn and J. Buikstra (eds.) *Human Identification*. Springfield, IL: Charles C. Thomas, pp. 253–277.

Sobrian S. K., Vaughn V. T., Ashe W. K., *et al.* (1997) Gestational exposure to loud noise alters the development and postnatal responsiveness of humoral and cellular components of the immune system in offspring. *Environmental Research* **73**:227–241.

Sontag L. W. and Comstock G. (1938) Striae in the bones of a set of monozygotic triplets. *American Journal of Diseases of Children* **56**:301–308.

Soren D. and Soren N. (eds.) (1999) *A Late Roman Villa and a Late Roman Infant Cemetery: Excavation in Poggio Gramignano Lugnano in Teverina*. Rome: l'Erma de Bretschneider.

Spence M. (1986) *The Excavation of the Keffer Site Burials*. London, Ontario: Museum of Indian Archaeology.

Spitz W. (1993) Thermal injuries. In W. U. Spitz (ed.) *Medicolegal Investigation of Death: Guidelines for the Application of Pathology to Crime Investigation*. Springfield, IL: Charles C. Thomas, pp. 413–443.

St Hoyme L. and Iscan M. (1989) Determination of sex and race: accuracy and assumptions. In M. Iscan and K. Kennedy (eds.) *Reconstruction of Life from the Skeleton*. New York: Alan R. Liss, pp. 53–93.

Starling S., Heller R. and Jenny C. (2002) Pelvic fractures in infants as a sign of physical abuse. *Child Abuse and Neglect* **26**:475–480.

Start H. and Kirk L. (1998) 'The bodies of friends': the osteological analysis of a Quaker burial ground. In M. Cox (ed.) *Grave Concerns: Death and Burial in England 1700–1850*. York, UK: Council for British Archaeology, pp. 167–177.

Steckel R. and Rose J. (2002) *The Backbone of History: Health and Nutrition in the Western Hemisphere*. Cambridge, UK: Cambridge University Press.

Steckel R., Sciulli P. and Rose J. (2002) A health index from skeletal remains. In R. H. Steckel and J. C. Rose (eds.) *The Backbone of History: Health and Nutrition in the Western Hemisphere*. Cambridge, UK: Cambridge University Press, pp. 61–93.

Steinbach H. L. (1966) Infections in bones. *Seminars in Roentgenology* **1**:337–369.

Steiniche T. and Hauge E. (2003) Normal structure and function of bone. In H. Yuehuei and K. Martin (eds.) *Handbook of Histology: Methods for Bone and Cartilage*. Totowa, NJ: Humana Press, pp. 59–72.

Stephenson L., Latham M. and Jansen A. (1983) *A Comparison of Growth Standards: Similarities between Harvard and Privileged African Children and Differences with Kenyan Rural Children*, Cornell International Nutrition Monograph Series. Ithaca, NY: Cornell University Press.

Stevenson D., Verter J., Fanaroff A., *et al.* (2000) Sex differences in outcomes of very low birthweight infants: the newborn male disadvantage. *Archives of Diseases in Childhood Fetal and Neonatal Edition* **83**:182–185.

Stevenson P. (1924) Age order of epiphyseal union in man. *American Journal of Physical Anthropology* **7**:53–93.

Steyn M. and Henneberg M. (1996) Skeletal growth of children from the Iron Age site at K2 (South Africa). *American Journal of Physical Anthropology* **100**:389–396.

Steyn M. and Henneberg M. (1997) Cranial growth in the prehistoric sample from K2 at Mapungubwe (South Africa) is population specific. *Homo* **48**:62–71.

Still G. (1912) *Common Disorders and Diseases of Childhood*. London: Froude.

Stini W. A. (1969) Nutritional stress and growth: sex difference in adaptive response. *American Journal of Physical Anthropology* **31**:417–426.

Stini W. A. (1975) Adaptive strategies of human populations under nutritional stress. In E. S. Watts, F. E. Johnston and G. W. Lasker (eds.) *Biosocial Interrelations in Population Adaptation*. Paris: Mouton, pp. 19–41.

Stloukal M. and Hanáková H. (1978) Die Länge der Längsknochen altslawischer Bevölkerungen – unter besonderer Berückichtigung von Wachstumsfragen. *Homo* **29**:53–69.

Stockwell E. (1993) Infant mortality. In K. Kiple (ed.) *The Cambridge World History of Human Disease*. Cambridge, UK: Cambridge University Press, pp. 224–229.

Stodder A. (1997) Subadult stress, morbidity and longevity in Latte Period populations on Guam, Mariana Islands. *American Journal of Physical Anthropology* **104**:363–380.

Stodder A. (2005) Treponemal infection in the prehistoric Southwest. In M. Powell and D. Cook (eds.) *The Myth of Syphilis: The Natural History of Treponematosis in North America*. Gainesville, FL: University of Florida Press, pp. 227–280.

Stone L. (1977) *The Family, Sex and Marriage in England 1500–1800*. London: Weidenfeld & Nicolson.

Stoodley N. (2000) From the cradle to the grave: age organisation and the early Anglo-Saxon burial rite. *World Archaeology* **31**:456–472.

Storey R. (1986) Perinatal mortality at pre-Columbian Teotihuacán. *American Journal of Physical Anthropology* **69**:541–548.

Storey R. (1988) Prenatal enamel defects in Teotihuacán and Copan. *American Journal of Physical Anthropology* **75**:275–276.

Strand Vidarsdottir U. (2003) MorFIDS: morphometric forensic identification of sub-adults. *American Journal of Physical Anthropology* Suppl. **36**:202.

Strand Vidarsdottir U., O'Higgins P. and Stringer C. (2002) A geometric morphometric study of regional differences in the ontology of the modern human facial skeleton. *Journal of Anatomy* **201**:211–229.

Stroud G. and Kemp R. (eds.) (1993) *Cemetery of St Andrew Fishergate*. London: Council for British Archaeology.

Stuart-Macadam P. L. (1987) A radiographic study of porotic hyperostosis. *American Journal of Physical Anthropology* **74**:511–520.

Stuart-Macadam P. L. (1988) Rickets as an interpretative tool. *Journal of Paleopathology* **2**:33–42.

Stuart-Macadam P. L. (1989) Nutritional deficiency diseases: a survey of scurvy, rickets and iron-deficiency anemia. In M. Y. Iscan and K. A. R. Kennedy (eds.) *Reconstruction of Life from the Skeleton*. New York: Alan R. Liss, pp. 201–222.

Stuart-Macadam P. L. (1991) Anemia in Roman Britain: Poundbury Camp. In H. Bush and M. Zvelebil (eds.) *Health in Past Societies: Biocultural Interpretations of*

Human Skeletal Remains in Archaeological Contexts, BAR (International Series) no. S567. Oxford, UK: Archaeopress, pp. 101–113.

Stuart-Macadam P. L., Glencross B. and Kricun M. (1998) Traumatic bowing deformities in tubular bones. *International Journal of Osteoarchaeology* **8**:252–262.

Suckling G. and Thurley D. C. (1984) Developmental enamel defects of enamel: factors influencing their macroscopic appearance. In R. Fearnhead and S. Suga (eds.) *Tooth Enamel*. New York: Elsevier, pp. 357–362.

Sundick R. (1972) Human skeletal growth and dental development as observed in the Indian Knoll population. Ph.D. thesis, University of Toronto, Canada.

Sundick R. I. (1978) Human skeletal growth and age determination. *Homo* **29**:228–249.

Sutter R. (2003) Nonmetric subadult skeletal sexing traits. I. A blind test of the accuracy of eight previously proposed methods using prehistoric known-sex mummies for Northern Chile. *Journal of Forensic Sciences* **48**:20–25.

Swanson J. (1990) Childhood and childrearing in status sermons by later thirteenth century friars. *Journal of Medieval History* **16**:309–331.

Swedlund A. and Armelagos G. (1969) Une recherche en paléo-démographie: la Nubie soudanaise. *Annales: Économies, Sociétés, Civilisations* **24**:1287–1298.

Swedlund A. and Armelagos G. (1976) *Demographic Anthropology*. Dubuque, IA: William C. Brown.

Sweeney E. A., Saffir A. J. and De Leon R. (1971) Linear hypoplasia of deciduous incisor teeth in malnourished children. *American Journal of Clinical Nutrition* **24**:29–31.

Swerdloff B. A., Ozonoff M. B. and Gyepes M. T. (1970) Late recurrence of infantile cortical hyperostosis (Caffey's disease). *American Journal of Roentgenology* **108**:461–467.

Tanaka T., Cohen P., Clayton P., *et al.* (2002) Diagnosis and management of growth hormone deficiency in childhood and adolescence. II. Growth hormone treatment in growth hormone deficient children. *Growth Hormone and IGF Research* **12**:323–341.

Tanguay R., Buschang P. and Demirjian A. (1986) Sexual dimorphism in the emergence of deciduous teeth: its relationship with growth components in height. *American Journal of Physical Anthropology* **69**:511–515.

Tanner J. M. (1978) *Foetus into Man*. London: Open Books Publications Ltd.

Tanner J. M. (1981) Catch-up growth in man. *British Medical Bulletin* **37**:233–238.

Tanner J. M. (1986) Growth as a target-seeking function. In F. Falkner and J. M. Tanner (eds.) *Human Growth: A Comprehensive Treatise*, vol. 1. London: Plenum Press, pp. 167–179.

Tanner J. M. and Eveleth P. B. (1976) Urbanisation and growth. In G. A. Harrison and J. B. Gibson (eds.) *Man in Urban Environments*. Oxford, UK: Oxford University Press, pp. 144–166.

Tanner J. M., Whitehouse R., Marshall W., Healy J. and Goldstein H. (1975) *Assessment of Skeletal Maturity and Prediction of Adult Height*. London: Academic Press.

Tardieu A. (1860) Étude médico-légale sur les services et mauvais traitements exercés sur enfants. *Annales d'Hygiène Publique et de Médecine Légale* **13**:361–398.

Teegan W.-R. (2004) Hypoplasia of the tooth root: a new unspecific stress marker in human and animal palaeopathology. *American Journal of Physical Anthropology* Suppl. **38**:193.

Teivens A., Mörnstad H., Norén J. G. and Gidlund E. (1996) Enamel incremental lines as recorders for disease in infancy and their relation to the diagnosis of SIDS. *Forensic Science International* **81**:175–183.

Teklali Y., El Alami Z., El Madhi T., Gourinda H. and Miri A. (2003) Peripheral osteoarticular tuberculosis in children: 106 case-reports. *Joint, Bone, Spine* **70**:282–286.

Telkka A., Palkama A. and Vertama P. (1962) Prediction of stature from radiographs of long bones in children. *Journal of Forensic Sciences* **7**:474–476.

Teschler-Nicola M. (1997) Differential diagnosis of tuberculosis: the diagnostic value of endocranial features. Poster presented at the *International Congress on the Evolution and Palaeoepidemiology of Tuberculosis*, Szeged, Hungary.

Teschler-Nicola M., Gerold F. and Prodinger W. (1998) Endocranial features in tuberculosis. Paper presented at the *12th European Meeting of the Paleopathology Association*, Prague.

Thacher T. D., Fischer P., Pettifor J., *et al.* (2000) Radiographic scoring method for the assessment of the severity of nutritional rickets. *Journal of Tropical Pediatrics* **46**:132–139.

Thacher T. D., Fischer P. R., Pettifor J. M., Lawson J. O. and Isichei C. O. (1999) A comparison of calcium, vitamin D, or both for nutritional rickets in Nigerian children. *New England Journal of Medicine* **341**:563–568.

Thompson B. (1984) Infant mortality in nineteenth-century Bradford. In R. Woods and J. Woodward (eds.) *Urban Disease and Mortality in Nineteenth Century England*. London: Batsford, pp. 120–147.

Thompson G., Anderson D. and Popovich F. (1975) Sexual dimorphism in dentition mineralisation. *Growth* **39**:289–301.

Thomson A. (1899) The sexual differences of the foetal pelvis. *Journal of Anatomy and Physiology* **33**:359–380.

Thorpe I. (1996) *The Origins of Agriculture in Europe*. London: Routledge.

Thorpe I. (2003) Anthropology, archaeology, and the origin of warfare. *World Archaeology* **35**:145–165.

Tillier A.-M., Arensburg B., Duday H. and Vandermeersch B. (2001) Brief communication: an early case of hydrocephalus: the Middle Paleolithic Qafzeh 12 child. *American Journal of Physical Anthropology* **114**:166–170.

Timby J. R. (1996) *The Anglo-Saxon Cemetery at Empingham II, Rutland: Excavations carried out between 1974 and 1975*. Oxford, UK: Oxbow Books.

Titley K., Pytnn B., Chernecky R., *et al.* (2004) The Titanic disaster: dentistry's role in the identification of an 'unknown child'. *Journal of the Canadian Dental Association* **70**:24–28.

Tobias P. (1958) Studies on the occipital bone in Africa. *South African Journal of Medical Science* **23**:135–146.

Tocheri M. and Molto J. (2002) Ageing fetal and juvenile skeletons from Roman period Egypt using basiocciput osteometrics. *International Journal of Osteoarchaeology* **12**:356–363.

Tocheri M., Dupras T. and Molto J. (2001) In or out of the womb? The analysis and interpretation of fifteen fetal skeletons from Roman Period Egypt. *American Journal of Physical Anthropology* Suppl. **32**:150.

Todd A., Baker R. and Panter-Brick C. (1996) Physical growth of homeless boys in Nepal. *Journal of the National Medical Association* **34**:152–158.

Todd T. (1930) The anatomical features of epiphyseal union. *Child Development* 1:186–194.

Tomkins R. L. (1996) Human population variability in relative dental development. *American Journal of Physical Anthropology* 99:79–102.

Tooley M. (1983) *Abortion and Infanticide*. Oxford, UK: Clarendon Press.

Torwalt C. R., Balachandra A., Youngson C. and de Nanassy J. (2002) Spontaneous fractures in the differential diagnosis of fractures in children. *Journal of Forensic Sciences* 47:1–5.

Treuta J. (1959) The three types of haematogenous osteomyelitis. *Journal of Bone and Joint Surgery* 41–B:671–680.

Trinkaus E. (1977) The Alto Salaverry child: a case of anemia from the Peruvian Preceramic. *American Journal of Physical Anthropology* 46:25–28.

Trinkaus E., Formicolal V., Svoboda J., Hillson S. and Holliday T. (2001) Dolni Vestonice 15: pathology and persistence in the Pavlovian. *Journal of Archaeological Science* 28:1291–1308.

Trotter M. (1971) The density of bones in the young skeleton. *Growth* 35:221–231.

Turlington E. G. (1970) Chronic sclerosing nonsuppurative osteomyelitis. *Transactions of the 4th International Conference of Oral Surgery* 6:120–124.

Ubelaker D. H. (1974) *Reconstruction of Demographic Profiles from Ossuary Skeletal Samples: A Case Study from the Tidewater Potomac*. Washington, DC: Smithsonian Institution Press.

Ubelaker D. H. (1989) *Human Skeletal Remains: Excavation, Analysis, Interpretation*. Washington, DC: Taraxacum Press.

Ulijaszek S. J. (1990) Nutritional status and susceptibility to infectious disease. In G. A. Harrison and J. C. Waterlow (eds.) *Diet and Disease in Traditional and Developing Societies*. Cambridge, UK: Cambridge University Press, pp. 137–154.

Ulijaszek S. J. (1998) Measurement error. In S. Ulijaszek, F. E. Johnston and M. A. Preece (eds.) *The Cambridge Encyclopedia of Human Growth and Development*. Cambridge, UK: Cambridge University Press, p. 28.

Ulijaszek S. J., Johnston F. E. and Preece M. A. (eds.) (1998) *The Cambridge Encyclopedia of Human Growth and Development*. Cambridge, UK: Cambridge University Press.

Ulizzi L. and Zonta L. (2002) Sex differential patterns in perinatal deaths in Italy. *Human Biology* 74:879–888.

Van Gennep A. (1960) *The Rites of Passage*. Chicago, IL: University of Chicago Press.

Van Gerven D., Hummert J. and Burr D. (1985) Cortical bone maintenance and the geometry of the tibia in prehistoric children from Nubia's Batn El Hajar. *American Journal of Physical Anthropology* 66:275–280.

Van Leeuwen B., Hartel R., Jansen H., Kamps W. and Hoekstra H. (2003) The effect of chemotherapy on the morphology of the growth plate and metaphysis of the growing skeleton. *European Journal of Surgical Oncology* 29:49–58.

Visser E. P. (1998) Little waifs: estimating child body size from historic skeletal material. *International Journal of Osteoarchaeology* 8:413–423.

Vögele J. P. (1994) Urban infant mortality in Imperial Germany. *Social History of Medicine* 7:401–426.

Von Endt D. W. and Ortner D. J. (1984) Experimental effects of bone size and temperature on bone diagenesis. *Journal of Archaeological Science* 11:247–253.

Von Fehling H. (1876) Die Form des Beckens biem Fötus und neugeborenen und ihre Beziehung zu der beim Erwachsenen. *Archiv der Gynäkologie Berlin* **10**:1–80.

Waaler P. (1983) Anthropometric studies in Norwegian children. *Acta Paediatrica Scandinavica* **308**:1–41.

Waldron T. (1988) The human remains. In V. Evison (ed.) *An Anglo-Saxon Cemetery at Great Chesterford, Essex*. York, UK: Council for British Archaeology pp. 52–66.

Waldron T. (1993) The human remains from the Royal Mint Site (MIN86), unpublished report. London: Museum of London Archive Report HUM/07/93.

Waldron T. (2000) Hidden or overlooked? Where are the disadvantaged in the skeletal record? In J. Hurbert (ed.) *Madness, Disability and Social Exclusion*. London: Routledge, pp. 29–45.

Waldron T., Taylor G. and Rudling D. (1999) Sexing of Romano-British baby burials from the Beddingham and Bignor villas. *Sussex Archaeological Collections* **137**:71–79.

Walker D. and Walker G. (2002) Forgotten but not gone: the continuing scourge of congenital syphilis. *Lancet Infectious Diseases Journal* **2**:432–436.

Walker P. (1969) The linear growth of long bones in Late Woodland Indian children. *Proceedings of the Indian Academy of Science* **78**:83–87.

Walker P. (1986) Porotic hyperostosis in a marine-dependent California Indian population. *American Journal of Physical Anthropology* **69**:345–354.

Walker P. (1997) Skeletal evidence for child abuse: a physical anthropological perspective. *Journal of Forensic Sciences* **42**:196–207.

Walker P., Johnson J. and Lambert P. (1988) Age and sex biases in the preservation of human skeletal remains. *American Journal of Physical Anthropology* **76**:183–188.

Wall C. E. (1991) Evidence of weaning stress and catch-up growth in the long bones of a Central California Amerindian sample. *Annals of Human Biology* **18**:9–22.

Walls T. and Shingadia D. (2004) Global epidemiology of pediatric tuberculosis. *Journal of Infection* **48**:13–22.

Wapler U., Crubézy E. and Schultz M. (2004) Is cribra orbitalia synonymous with anemia? Analysis and interpretation of cranial pathology in Sudan. *American Journal of Physical Anthropology* **123**:333–339.

Warren M. W. (1999) Radiographic determination of developmental age in fetuses and stillborns. *Journal of Forensic Sciences* **44**:708–712.

Warren M. W., Holliday T. and Cole T. (2002) Ecogeographical patterning in the human fetus. *American Journal of Physical Anthropology* Suppl. **34**:161.

Warwick R. (1986) Anne Mowbray: skeletal remains of a medieval child. *London Archaeology* **5**:176–179.

Waters A. and Katzenberg A. (2004) The effects of growth velocity on stable nitrogen isotope ratios in subadult long bones. *American Journal of Physical Anthropology* Suppl. **38**:204.

Watt M., Lunt D. and Gilmour W. (1997) Caries prevalence in the deciduous dentition of a medieval population from the south-west of Scotland. *Archives of Oral Biology* **42**:811–820.

Watts D. J. (1989) Infant burials and Romano-British Christianity. *Archaeological Journal* **146**:372–383.

Way J. (2003) Fetal and subadult age estimation using the os temporale pars petrosa: accuracy of quantitative and qualitative criteria. *American Journal of Physical Anthropology* Suppl. **36**:221.

Weaver D. (1980) Sex differences in the ilia of a known sex and age sample of fetal and infant skeletons. *American Journal of Physical Anthropology* **52**:191–195.

Weinberg E. D. (1974) Iron susceptibility to infectious disease. *Science* **184**:952–956.

Weinberg E. D. (1992) Iron withholding in prevention of disease. In P. Stuart-Macadam and S. Kent (eds.) *Diet, Demography and Disease: Changing Perspectives on Anemia.* New York: Aldine de Gruyter, pp. 105–150.

Weinberg S., Putz D., Mooney M. and Siegal M. (2002) Nonmetric population variation in perinatal human skulls. *American Journal of Physical Anthropology* Suppl. **34**:162.

Weiss K. (1973) *Demographic Models for Anthropology*, Memoir no. 27. Washington, DC: Society for American Archeology.

Wells C. (1964) *Bones, Bodies and Disease.* London: Thames & Hudson.

Wells C. (1975a) Ancient obstetric hazards and female mortality. *Bulletin for the New York Academy of Medicine* **51**:1235–1249.

Wells C. (1975b) Prehistoric and historical changes in nutritional diseases and associated conditions. *Progress in Food and Nutritional Science* **1**:729–779.

Wells C. (1978) A medieval burial of a pregnant woman. *Practitioner* **221**:442–444.

Wells C. (1996) Human bones. In V. Evison and P. Hill (eds.) *Two Anglo-Saxon Cemeteries at Beckford, Hereford and Worcester.* York, UK: Council for British Archaeology, pp. 41–62.

West S. (1888) Acute periosteal swellings in several young infants of the same family, probably rickety in nature. *British Medical Journal* **1**:856–857.

Whistler D. (1645) *De Morbo Puerli Anglorum, Quem Patrio Ideiomate Indigenae Vocant "The Rickets".* Batavorum: Lugduni.

White C., Longstaffe F. and Law K. (2004) Exploring the effects of environment, physiology and diet on oxygen isotope ratios in ancient Nubian bones and teeth. *Journal of Archaeological Science* **31**:233–250.

White C. and Schwartz H. (1994) Temporal trends in stable isotopes for Nubian mummy tissues. *American Journal of Physical Anthropology* **93**:165–187.

White T. D. (1978) Early hominid enamel hypoplasia. *American Journal of Physical Anthropology* **49**:79–84.

White W. (1988) The human bones: skeletal analysis. In W. J. White (ed.) *Skeletal Remains from the Cemetery of St Nicholas Shambles, City of London.* Over Wallop, UK: London and Middlesex Archaeological Society, pp. 28–55.

Whittaker D. and MacDonald D. (1989) *A Colour Atlas of Forensic Dentistry.* London: Wolfe.

Whittle A. (1996) *Europe in the Neolithic: The Creation of New Worlds.* Cambridge, UK: Cambridge University Press.

Wicker N. (1998) Selective female infanticide as partial explanation for the dearth of women in Viking Age Scandinavia. In G. Halstall (ed.) *Violence and Society in the Early Medieval West.* Woodbridge, UK: Boydell Press, pp. 205–221.

Wiggins R. (1991) Porotic hyperostosis, cribra orbitalia, enamel hypoplasia, periosteal reaction and metopism: a correlation of their prevalence and an assessment of the nature of porotic hyperostosis in three British archaeological populations. M.Sc. dissertation, University of Bradford, Bradford, UK.

Wiggins R. (1996) Skeletal stress indicators and long bone length for age in the subadult population from St. Peter's Church, Barton on Humber: an ongoing study. *Proceedings of the 11 European Members' Meeting of the Paleopathology Association*, Maastricht, The Netherlands.

Wilber J. and Thompson G. (1998) The multiply injured child. In N. Green and M. Swiontkowski (eds.) *Skeletal Trauma in Children*. Philadelphia, PA: W. B. Saunders, pp. 71–102.

Wiley A. S. and Pike I. L. (1998) An alternative method for assessing early mortality in contemporary populations. *American Journal of Physical Anthropology* **107**:315–330.

Wilkie L. (2000) Not merely child's play: creating a historical archaeology of children and childhood. In J. Sofaer Derevenski (ed.) *Children and Material Culture*. London: Routledge, pp. 100–113.

Willems G., Van Olmen A., Spiessens B. and Carels C. (2001) Dental age estimation in Belgian children: Demirjian's technique revisited. *Journal of Forensic Sciences* **46**:893–895.

Williams J., White C. and Longstaffe F. (2002) Trophic level and macronutrient shift effects associated with the weaning process in the Maya Postclassic. *American Journal of Physical Anthropology* Suppl. **34**:165.

Williams N. and Galley C. (1995) Urban–rural differentials in infant mortality in Victorian England. *Population Studies* **49**:401–420.

Williams P. L. and Warwick R. (1980) *Gray's Anatomy*, 36th edn. Edinburgh, UK: Churchill Livingstone.

Wilson L. G. (1975) The clinical definition of scurvy and the discovery of vitamin C. *Journal of the History of Medicine* **30**:40–60.

Wilson L. G. (1986) The historical riddle of milk-borne scarlet fever. *Bulletin of the History of Medicine* **60**:321–342.

Winnik H. and Horovitz M. (1961–2) The problem with infanticide. *British Journal of Criminology* **2**:40–52.

Wood J. W., Milner G. R., Harpending H. C. and Weiss K. M. (1992) The osteological paradox: problems of inferring prehistoric health from skeletal samples. *Current Anthropology* **33**:343–370.

Wood L. (1996) Frequency and chronological distribution of linear enamel hypoplasia in a North American colonial skeletal sample. *American Journal of Physical Anthropology* **100**:247–259.

Working Group (1994) *Weaning and the Weaning Diet*. London: Her Majesty's Stationery Office.

World Health Organisation (1986) Use and interpretation of anthropomorphic indicators of nutritional status. *Bulletin of the World Health Organisation*. **64**:929–941.

Worthen C., Flinn M., Leone D., Quinlan R. and England B. (2001) Parasite load, growth, fluctuating asymmetry, and stress hormone profiles among children in a rural Caribbean village. *American Journal of Physical Anthropology* Suppl. **32**:167.

Worthman C. M. and Panter-Brick C. (1996) Street versus home, urban versus rural, disadvantaged versus secure: contrasting lifestyles and well-being status of Nepali children. Paper presented at the *Symposium on Child Stress in Natural Settings: Psychobiological Studies of Children in Diverse Cultures, Biannual Meetings of the International Study of Behavioural Development*, Quebec City, Quebec, Canada.

Wright L. E. (1990) Stresses of conquest: a study of Wilson bands and enamel hypoplasias in the Maya of Lamanai, Belize. *American Journal of Human Biology* **2**:25–35.

Wright L. E. (1997) Intertooth patterns of hypoplasia expression: implications for childhood health in the Classic Maya Collapse. *American Journal of Physical Anthropology* **102**:233–247.

Wright L. E. (1998) Stable carbon and oxygen isotopes in human tooth enamel: identifying breastfeeding and weaning in prehistory. *American Journal of Physical Anthropology* **106**:1–18.

Wright L. E. and Schwartz H. (1997) Weaning in prehistory: a new isotopic method. *American Journal of Physical Anthropology* Suppl. **24**:246.

Wright L. E. and Schwartz H. P. (1999) Correspondence between stable carbon, oxygen and nitrogen isotopes in human tooth enamel and dentine: infant diets at Kaminaljuyú. *Journal of Archaeological Science* **26**:1159–1170.

Wrigley E. (1977) Births and baptisms: the use of Anglican baptism registers as a source of information about numbers of births in England before the beginning of civil registration. *Population Studies* **31**:281–312.

Wu T., Buck G. and Mendola P. (2003) Blood lead levels and sexual maturation in U.S. girls: the third national health and nutrition examination survey, 1988–1994. *Environmental Health Perspectives* **111**:737–741.

Yamamoto T., Uchili R., Kojima T., *et al.* (1998) Maternal identification from skeletal remains of an infant kept by the alleged mother for 16 years with DNA typing. *Journal of Forensic Sciences* **43**:701–705.

Y'Edynak G. (1976) Long bone growth in Western Eskimo and Aleut skeletons. *American Journal of Physical Anthropology* **45**:569–574.

Index

(index)

suicide, 15–16
Sunghir, Russia, double burial, 30
syphilis
  aetiology, 151–152
  archaeological cases, *155*, 157–159
  dental stigmata, 156–157, 158
  in modern populations, 152
  pathogenesis, 152–157
  syphilitic snuffles, 155

taphonomy
  bias, 22
  definition, 23–26
Taunton, Somerset, 32
Teotihuacan (Tlajjinga), Mexico, 82, 87
third molar
  calcification, 39
  eruption, 41
time since death, 23, 24
Titanic disaster (1912), 18
Tophet (Punic Carthage), 92
toys, 8
  and play, 9
  ivory dolls, Yverdon-les-Bains, 9
  miniature artefacts, 10
  Ovčarovo, Bulgaria, 9, *9*
  Platia Magoula Zarkou, 9
trauma, *see also* child abuse
  birth trauma, 173–175
  blunt force, 15
  clavicular, 169
  cranial, 87, 95, 168, 169, 170, 171
  greenstick, 164, *164*, 164–165
  healing rates, 167, 168
  plastic deformation, 164, 165, 172
  recording, *166*, 172–173
  rib fractures, 17, 124, 172
  shortened limbs, 171, *171*, *172*, 173
  spinal, 169
  subtlety of lesions, 170
  types, 163–164
trepanations, 170
Trotter Collection, Cleveland, 53
tuberculosis, 12, 31
  aetiology, 146
  diagnosis, *150*, 149–151
  endocranial lesions in, 143, 150–151
  history, 148–149
  and milk farms, 103

osteomyelitis, 141
pathogenesis, 146–148
typhoid, 141

Uley, Gloucestershire, 91
unbaptised, *see* non-baptised
under-representation, 11, 22, 33, 86
  in Anglo-Saxon cemeteries, 23, 86
  in Roman cemeteries, 86
Upper Palaeolithic burials, 30

Van Gennep, 8
Vedbaek, Scandinavia, 31
vitamin C deficiency, *see* scurvy
vitamin D deficiency, *see* rickets *and*
  osteomalacia
Vlasac, Serbia, 31
Voegtly cemetery, Pittsburgh, 22

Waco (Branch Davidian Compound), 16
Watchfield cemetery, Oxfordshire, 27
weaning
  age of, 101, 102–103, 115–117, 118
  condensed milk, 102–103
  cows' milk, 99–100, 103
  definition, 99
  dental enamel hypoplasias, 106–107
  dental wear, 42, 118
  diet, 22, 102, 117, 118
  feeding vessels, 102
  and isotope analysis, 12, 115–119
  'weanling's dilemma', 63, 100
Westerhus, Sweden, 87
West Heslerton, North Yorkshire, 23
wet-nursing, 33, 101–102
  and child abuse, 176
  and syphilis, 152
Wharram Percy, North Yorkshire, 50, 87, 94,
  116
Whitehawk Neolithic Camp, 34
whooping cough, and rickets, 121
Wilson bands, 105, 107
Wimberger's sign, 154, 159
wormian bones, and personal identification, 16
woven bone, *see* fibre bone

Yasmina, Carthage (child cemetery), 32, 92

zinc (and lactation), 37